A

[signature: Philip E. Lilienthal]

BOOK

The Philip E. Lilienthal imprint
honors special books
in commemoration of a man whose work
at University of California Press from 1954 to 1979
was marked by dedication to young authors
and to high standards in the field of Asian Studies.
Friends, family, authors, and foundations have together
endowed the Lilienthal Fund, which enables UC Press
to publish under this imprint selected books
in a way that reflects the taste and judgment
of a great and beloved editor.

The publisher gratefully acknowledges the generous support of the Philip E. Lilienthal Asian Studies Endowment Fund of the University of California Press Foundation, which was established by a major gift from Sally Lilienthal.

Fabricating Consumers

ASIA: LOCAL STUDIES/GLOBAL THEMES

Jeffrey N. Wasserstrom, Kären Wigen, and Hue-Tam Ho Tai, Editors

Fabricating Consumers

The Sewing Machine in Modern Japan

Andrew Gordon

UNIVERSITY OF CALIFORNIA PRESS

Berkeley · Los Angeles · London

University of California Press, one of the most distinguished university presses in the United States, enriches
lives around the world by advancing scholarship in
the humanities, social sciences, and natural sciences.
Its activities are supported by the UC Press Foundation
and by philanthropic contributions from individuals
and institutions. For more information, visit
www.ucpress.edu.

University of California Press
Berkeley and Los Angeles, California

University of California Press, Ltd.
London, England

Library of Congress Cataloging-in-Publication Data

Gordon, Andrew, 1952–
    Fabricating consumers : the sewing machine in
  modern Japan / Andrew Gordon. — 1st ed.
        p.   cm. — (Asia: local studies/global themes ; 19)
    Includes bibliographical references and index.
    ISBN 978-0-520-26785-5 (cloth : alk. paper)
    1. Singer Sewing Machine Company—History.
2. Sewing-machine industry—United States—History—
20th century.   3. Clothing trade—Japan—History—
20th century.   4. Consumers—Japan—History—20th
century.   I. Title.
    HD9971.5.S484S565   2011
    338.7'64620440952—dc23   2011019065

20   19   18   17   16   15   14   13   12   11
10   9   8   7   6   5   4   3   2   1

Contents

Illustrations

Preface

Over the past decade, I have been working on this history of the selling, buying, and using of the household sewing machine in Japan. As I pursued this project, I was impressed to learn in conversations with numerous friends and colleagues how common it had been to own a Singer sewing machine, at least for middle-class city-dwellers in Japan of the 1930s or later, the generation of my colleagues' parents. And as I began to present preliminary thoughts and results of my research to academic and broader public audiences, I was further impressed by the emotional resonance the sewing machine held for so many.[1]

One colleague, for example, wrote in a letter after he heard me speak, "Listening to your talk the other night, I felt as if I was hearing my own history. . . . My mother supplemented the family income as a dressmaker. She made clothes on order from neighbors or acquaintances and also sold through a local retail clothing store. . . . If I am permitted to exaggerate slightly, the sewing machine is an object that links to memories of my mother a bit like the sled 'Rosebud' in the film *Citizen Kane*."[2]

Another colleague—I cannot recall who it was—came up to me after a talk and said that, influenced subconsciously, he was sure, by knowledge of my upcoming talk, the previous night he had dreamed of his deceased mother sitting at her sewing machine. In more than twenty years of writing and speaking on the history of (mainly male) industrial workers, I received no such emotionally linked reactions. I began to realize this was a topic worth pursuing seriously in part because of the long-lived meanings the sewing machine held not only for those who used it but also for the members of their families. My initial impression was that in a gender-divided landscape of memory, the sewing machine was of particular resonance for the sons of sewing mothers, but this proved incorrect. The breakdown of reactions simply reflected the fact that the majority of academics in Japan, and thus most of those in the audience at talks, are men. The sewing machine certainly carried great meaning for women as well as for men, both those who used it and those who watched their family members do so. I want to thank all those who have shared these recollections with me over the years, as well as those who offered advice, research assistance, or financial support for this project.

• • •

I much appreciate the feedback from audiences over the years in talks at Dōshisha University, Hōsei University, Hokkaido University, Hitotsubashi University, Kyoto University, Kokugakuin University, Kobe University, Kansai University, Shōwa Women's University, Tachibana University, and Tokyo University in Japan, as well as Columbia, Duke, Harvard, the London School of Economics, Stanford, Oxford, and Yale in the United States and Britain.

The Area Studies Faculty at Tokyo University's Komaba Campus and the Institute for Social Science at the Hongo Campus served as hospitable and lively home bases for a year of research in 2002–03, and I am most grateful to my hosts there, Mitani Hiroshi at Komaba, and Nitta Michio at the Institute of Social Science.[3] Over the years I benefited from research grants from the Asia Center and Reischauer Institute of Japanese Studies at Harvard, and the Lee and Juliet Folger Family Fund. A fellowship from the Radcliffe Institute in the year 2007–8 was invaluable in giving me the time and the support of a stimulating intellectual community to allow me to pull my ideas together in book form. I am grateful to the members of our Radcliffe Institute writing group, Catherine Lutz, Timothy Rood, and Frances Kissling for incisive comments on early chapters, and to Sheldon Garon and Sarah A. Gordon for careful readings and feedback on the penultimate draft. I owe special thanks to Andrew Godley for comments on various draft chapters and for his help in understanding Singer's business records.

For comments and advice in discussions over the years, I also thank Barbara Burman, Penelope Francks, Carol Gluck, Hirota Masaki, Janet Hunter, Geoffrey Jones, Kamiya Akiyoshi, Kimura Kenji, Kobayashi Norio, Koizumi Kazuko, Ian Miller, Nakamura Naofumi, Narita Ryūichi, Ohmaki

Kunichika, Suzuki Jun, Kären Wigen, Ann Waswo, and Yoshimi Shunya. For help in finding and offering access to documents, I am grateful to Abe Kenji of the *Nihon Mishin Times,* the staff at the Edo-Tokyo Museum and the Nature and Science Museum of the Tokyo University of Agriculture and Technology. For advice on comparative time-study data, thanks to Herrick Chapman, Nancy Cott, Jonathan Gershuny, Maya Jasanoff, and Mary Lewis. Not all the world's information is yet available via keyword searches of the Internet or databases, and I wish to thank several research assistants for tremendous help in scouring various sources to discover references to sewing or sewing machines at libraries in Tokyo and Cambridge—in some cases the old-fashioned way, page by page: June Hwang, Matsuda Haruka, Song Byongkwon, Toyoda Maho, Jennifer Yum. Jeremy Yellen worked long and hard tracking down illustrations and securing permissions. I am grateful to Margot Chamberlain for help in preparing the manuscript in the final stages, to Reed Malcolm, Kalicia Pivirotto, and Jacqueline Volin of the University of California Press for shepherding the manuscript into its book form, and to series editors Hue-Tam Tai, Jeffrey Wasserstrom, and Kären Wigen for their support. My family, in particular my wife Yoshie, offered precious support as well as diversion from the obsessive nature of my academic pursuits. For this I am deeply grateful.

Finally, of course, I remain responsible for any errors of omission or commission.

Introduction

More than a decade ago, while researching for a book on the postwar Japanese labor movement, I ran across a surprising fact of daily life that lodged in my mind. The decision to look into the history of the sewing machine and eventually to write this book began with an attempt to make sense of this datum: married women in 1950s Japan, each and every day, devoted more than two hours to sewing. The path from surprise and curiosity toward a set of worthwhile historical questions has been meandering. New concerns emerged along the way, while the matter of time spent at needlework lurked unresolved in the background. Its importance should become clear over the course of this study, but we can begin with its initial appearance as puzzle and provocation.

The fact in question was published in *The History of Labor in Kawasaki,* among the finest of the exhaustively prepared institutional or local histories for which Japanese government organizations and municipalities are well known. The compilers excerpted a 1951 social survey of working-class families conducted in the densely populated Kawasaki-Yokohama industrial belt adjacent to Tokyo. The eighty-seven wives of factory laborers who were not employed outside the home spent an average of 139 minutes each day sewing, not far behind the 179 minutes for cooking and far exceeding the time spent on any other daily activity.[1]

More than two hours every day for sewing? What were these women doing? I conjectured that hand-sewing accounted for this use of their time, a sign that impoverished, working-class families remained so cut off from the middle-class consumer economy that they made their own clothes and repaired them until they literally fell apart. But a colleague suggested there might be more to the story.[2] Having read that sewing machines had become popular in Japan by this time, she wondered whether all this needlework was in fact being done by hand. If not, its relationship to the consumer and industrial economy would take on a different meaning.

I tried without success to find the full text of the original survey to learn whether these women were stitching by hand or machine, but in the process I unearthed several similar studies. The 1950s were years when national and local government agencies, the labor movement, and the small cadre of labor scientists who had been concerned with the social impact of capitalism and factory labor for decades were energetically tracking the daily lives of working people. The most careful and detailed survey, conducted by the Women and Minors Bureau of the Ministry of Labor in 1952, looked at time-use among four hundred urban, working-class families in the Tokyo area. It found that a woman married to a factory laborer and not employed outside the home devoted *even more time,* 180 minutes each weekday, to sewing. This was virtually identical to the time devoted to preparing three meals (181 minutes). To my surprise, it turned out that 37 percent of the surveyed families owned sewing machines.[3]

Other sources made it clear that such families in this era were putting their machines to a wide variety of uses, including the fashioning of Western-style garments following the latest patterns featured in women's magazines. A sewing machine at the time cost roughly double the monthly wage of a typical office worker. Already, in 1952, many members of Japan's working class as defined by government social statisticians were able to purchase and use this expensive and widely desired "consumer durable" good. I enclose "consumer" good in scare quotes to highlight the fact that the sewing machine was in two ways androgynous. It was sold by men, but bought and used in the home by women. And it was both an object of consumer desire and a producer good used to make clothing not only for the family but for sale to neighbors or piece-good brokers. I began to realize that by studying the spread of the sewing machine, I could learn something of value about the social transformations of the twentieth century that turned Japan into a society in which the great majority understood themselves to be members of a modern middle class, defined not only by where they worked but also by what they bought and how they bought it.

This initial foray into social surveys made it clear that these long hours of sewing included significant time spent with machines, but it raised two important further questions. Were these hours really long compared to other places or times? Time-use studies in the United States and Western Europe detailed in chapter 7 do confirm that women in Japan spent a good deal more time at needlework than those in other modern societies. But why were these women—whether by hand or by machine—sewing so very much? As I explore in the chapters to follow, the answer rests in a dynamic interaction of local and global practice. Time-intensive home sewing has roots in the material culture of the Tokugawa era. From Meiji times through the 1950s, as the practices of dress and sewing shifted from Japanese to Western modes, the efforts of salesmen, the exhortations of civilian reformers and state officials, and the needs and desires of millions of women would together reinforce the time-consuming practice of fabricating clothing at home.

. . .

The sewing machine was so caught up in the development of capitalism worldwide that two of the most important critics of the modern condition wrote passionately about its place as an emblem of a new world. Karl Marx railed at length in *Capital* against this "decisively revolutionary machine, the machine which attacks in an equal degree the whole of the numberless branches of this sphere of production, dressmaking, tailoring, shoemaking, sewing, hat-making, and many others." The sewing machine on the one hand lowered the wages of the "better situated handicraftsmen" in these trades; its "overpowering competition crushes the weakest of the manual laborers. The fearful increase in death from starvation during the last 10 years in London runs parallel with the extension of machine sewing." On the other hand, Marx continued, this infernal machine collected "exclusively girls and young women" into narrow garrets where they endured long hours, cramped quarters, and "horrible" heat "due to the gas stoves used for warming the irons."[4]

Like Marx, Mahatma Gandhi "object[ed] to the 'craze' for machinery, not machinery as such. The craze is for what they call labour-saving machinery. Men go on 'saving labour' till thousands are without work and thrown on the open streets to die of starvation." But unlike Marx, when it came to the sewing machine, Gandhi qualified this indictment of machines and mass production: "I would make intelligent exceptions. Take the case of the Singer Sewing Machine. It is one of the few useful things ever invented and there is a romance about the device itself. Singer saw his wife labouring over the tedious process of sewing and seaming with her own hands, and simply out of his love for her he devised the sewing machine. . . . He, however, saved not only her labour but also the labour of everyone who could purchase a sewing machine."[5]

Gandhi was more than a little naive—or perhaps knowing but strategic—in his account of Isaac Merritt Singer's matrimonial commitment. The founder of the world's most successful sewing machine company was in his time a renowned philanderer, father to at least nineteen children by five wives or lovers. But Gandhi's view of the value of the sewing machine offers a nice counterpoint to that of Marx. While sharing fury at the devastating power of industrial capitalism, they point toward alternative understandings of the process later described as globalization. Marx saw capitalism as a leveling force of overwhelming power; it would transform the world everywhere in similar ways. Local industries and particularities would be destroyed or flattened as "all fixed, fast frozen relations, with their train of ancient and venerable prejudices and opinions, are swept away," and "all that is solid melts into air."[6] Only when the bourgeois revolution completed its destructive work would the proletariat emerge to usher in a better age. Gandhi, in contrast, believed that some ancient relations could and should survive. Focusing on homes and communities rather than workplaces, he thought it possible to fend off "industrialization on a mass scale" by respecting the "special conditions of India" and striving for "the revival of the village" as a self-contained unit. Some "modern machines and tools," most notably the sewing machine, he felt, could be deployed in a nonexploitative fashion.[7]

In more recent scholarly studies, issues of industrial capitalism and its worldwide diffusion have been analyzed as a process of globalization, with that key term understood by some, with Marx, as unrelentingly homogenizing in its impact and by others, with Gandhi, as notably varied in its actual or potential local unfoldings. Drawing on Foucault's conception of the "micro-power" of modern governmentality, Victoria de Grazia in her magisterial study of the trans-Atlantic history of America's "market empire" stands in the former camp. Although she takes care to note and to analyze the considerable "rancor, incomprehension and clash" in European reactions to the spread of this consumerist empire, she argues in the end that American capitalism in the twentieth century, especially the ethos and practice of mass merchandising and branding goods such as Singer's machine, constituted an "irresistible empire."[8] The "market empire" in this telling was a force of overwhelming power, which transformed and remade the civic and daily lives of twentieth-century

Europeans in America's image. Implicitly here, if Europe could not resist this empire, neither could the rest of the world.

In contrast, Arjun Appadurai has been among the most eloquent in arguing that central to the modern world's economic and cultural history has been a process in which the global and the local feed off each other. Appadurai understands "local" dynamically; unlike Gandhi, he is neither defending, nor restoring, nor seeking to recover for posterity local conditions understood to preexist and endure in some pristine form. Nonetheless, he does not see globalization as homogenization. He presents it, instead, as "a deeply historical, uneven, even *localizing* process." He notes that "at least as rapidly as forces from various metropolises are brought into new societies they tend to become indigenized in one or another way."[9]

In this book, through study of the selling, buying, and using of the sewing machine in Japan of the early-to-mid-twentieth century, I also understand globalization as a "localizing process," while recognizing its flattening power in many respects. This localization took place in both the realm of discourse—how people talked or wrote about this good—and in the world of social and economic practice. On the user, or demand, side, the focus is on home-based rather than factory users, although the story prominently includes commercial production from the home by seamstresses or dressmakers. On the supply side, the story starts with the Singer Sewing Machine Company and turns later to its Japanese competitors. I am less pessimistic than Marx on the matter of exploitation and less optimistic than Gandhi on the sewing machine's potential to usher in a more humane system of production and consumption. I am most impressed by the extraordinary variety of meanings and experiences attached to this object. Its simple name in the Japanese language, *mishin* (machine), made it an emblem of industrial technology. Famous on Broadway for its appearance in *Fiddler on the Roof* as the prize "child" of Motel the tailor, in Japan it shows up in locations ranging from the set of *Madame Butterfly* to memoirs of survival in the aftermath of World War II.[10] Found almost everywhere and leaving many traces, it allows us to study the modern transformation of daily life with its continuing harshness, its new opportunities, and its new imposition of discipline.

The sewing machine played a role in shaping the modern world. Historians consider Singer the world's first successful multinational company, and they have called the sewing machine a product that "set the paradigm of mass production and consumption."[11] But while I note such areas of impact or influence below, my primary concern is less with the sewing machine as prime mover in a causal chain and more with its value as a lens for examining varied aspects of the experience of modernity.

Certainly the selling of this good carried some homogenizing force. As described in chapter 2, Singer's practices of marketing and sales—its pioneering monthly installment sales and its network of stores and salesmen—allowed it to dominate the prewar Japanese market as it did in most of the world. These tactics varied remarkably little from country to country, a point of pride for Singer, whose managers boasted to the Japanese Foreign Ministry and wider public in 1932 that the "Singer system" was "universal throughout the world."[12] Yet one finds as well significant resistance to and negotiation over key elements of the Singer system, akin to what de Grazia describes as the "rancor, incomprehension and clash" found in Europe as it absorbed the shocks of the market empire. In a dramatic strike of the Japanese sales force against the Singer Sewing Machine Company in 1932–33, local employees begged to differ when expatriate managers defended the fairness and universality of their system. They literally smashed what they denounced as "Yankee capital" in a violent attack that destroyed the Yokohama headquarters. Angry salesmen then left Singer to join struggling local competitors. They eventually led the postwar domestic industry to overcome Singer at home and abroad with practices that drew on but in critical ways modified Singer's famed and "universal" system.

The story of the rise, fall, and adaptation of the Singer sales system in Japan, and especially the story of the manifold ways in which buyers, women in particular, made use and made sense of this good, offer a finely grained picture of globalization as a localizing process. As it did the world over, the sewing machine played a role generating selling practices to reach—and indeed create—the modern consumer, defined as someone engaged in a world of branded goods who satisfied both needs and desires by the regular and disciplined use of credit. These were global trends. But the sewing machine and new practices of sewing were connected in Japan to questions of dress and daily life that brought into relief problems of defining particularly "Japanese" and "Western" or "American" ways of life, and their relation to one another. We will see in chapter 3 that women and men in Japan defined their nation through cultural battles, including one over the merits of machine-made Western dress versus the hand-stitched kimono. Ironically, both sides to this debate, which took off in earnest in the 1920s, invoked the same modern virtues of efficiency and rationality. Even so, they reinforced an enduring dichotomy of the West versus Japan. The sewing machine in this and other ways was continually and diversely connected to a sense of being Japanese, American, or Western, or some problematic combination of these states of being.

A defining feature of the global history of the twentieth century in many places, including Japan, has been the movement of the consumer into the socioeconomic, political, and cultural spotlight. She edged toward center stage during the first half of the twentieth century, well into the years of total war, as the buyer and user of a plethora of household goods from sewing machines and pianos to bicycles, watches, and radios—goods that promised a life of progress, efficiency, and pleasure, that is, a modern life. In the postwar decades she and her "salaryman" husband and their children emerged in full glory as heroine and hero enjoying the bright new life of an ascendant middle class.

To be sure, in Japan as in many places, one finds a rise in consumption before one finds the culturally or politically prominent figure of the "consumer." Significant growth in the consumption of foodstuffs and other goods by a growing body of ordinary people in both cities and the countryside began in the eighteenth and nineteenth centuries.[13] In these same years one also finds anxious voices raising concern that all this getting and spending was out of hand. Such trends and attitudes continued, so one must not present the story from the late nineteenth century onward as a rupture with the past. Even so, one can identify significant differences in the place of consumption and consumers in Japan beginning around the turn of the twentieth century, so that the advent of the sewing machine—first sold by Singer directly to customers in exactly 1900—coincides nicely with the era of the modern consumer.

This consumer was a defining figure of a new age in several related ways. First, she participated in practices of getting and spending that were transnational or global in qualitatively different ways than in the past. European glassware and other goods imported through Nagasaki had been objects of curiosity and symbols of status in the Edo era. But by the twentieth century it was not only particular goods that were global or Western in origin and happily (or anxiously) understood to be so. Also identified with global modernity—of a particularly American origin—was a whole complex of practices from salesmanship, advertising, and consumer credit to mass formal education in home management. This was a complex designed to generate the desire and the possibility to possess branded goods that placed women and families either proudly and pleasurably in the middle class or made them anxiously seek to join it. Its transnational character meant that consuming, as noted above, was tightly wound up with defining what was "Japanese" about Japan and what was not. The constancy and depth of the concern to define Japanese-ness in consumer life contributed in significant ways to the process of inscribing national difference as a central element of the modern mindset.

A second defining feature of the consumer in this new age was her connection to a widespread concern over social division and inclusion. Again, the connection between social status and the possession and flaunting of certain goods that marked off divisions in society is hardly new to the modern era. What does appear different is how fear of class division as a major "social question"—for many it was *the* major issue of the day—was tied to a fear that Japan was on track to replicate a divisive and destructive Western experience. Most important, while exclusion from the new world of getting and spending might sharpen class alienation and antagonism, access to consumer goods also held the potential to happily answer this social question. In the chapters (especially 3, 5, and 7) to follow, I consider this social problem—the question of class—by examining the varied users of sewing machines, ranging from women engaged in homework for brokers or dressmaking for neighbors to those sewing "smart" outfits for their children. Most women bought their machines on time, and installment purchasing occasioned anxious debate globally and in Japan (hardly a debate only of times past) as to whether it might plunge overextended consumers into debt-ridden poverty or might instead cultivate disciplined, budget-minded, and upwardly striving subjects. We will see that, on balance, the purchase, ownership, and use of a good such as the sewing machine were steps in a complex process that gradually—if unevenly—integrated families of diverse economic means into a mass middle class.

I have deliberately used the female pronoun for the consumer because a third key element in the history of the modern consumer in Japan, as elsewhere, is the emergence of women into central and widely recognized social and economic roles, often based in the home but with consequences reaching beyond it. The efforts first of Singer and later of its competitors to promote a way of life requiring use of the sewing machine—and the manifold responses to these efforts—have left valuable traces, and their study offers numerous insights into the history of gender. Marketing appeals reflected and shaped a strikingly diverse array of female roles. Not surprisingly, sellers appealed to women as frugal and efficient household managers, but they also glorified women as seekers of independent livelihood and—even in wartime—as pleasure-seeking shoppers. One insight hardly limited to Japan, but not sufficiently recognized by historians of Japan, is that for a woman to be modern implicated her not in one female role but in a multitude of them.

In the chapters to follow, I follow this machine's journey into and through Japan, examining the varied transactions that linked buyers and users to sellers and makers and that connected people to the good itself, with attention also to the role of mass media, educators, and the state in mediating these transactions. Studied in this fashion, a small object allows a grounded approach to key aspects of modern history as experienced by people ranging from the exalted and powerful to the ordinary or powerless. Put to manifold uses with varied meanings, both a tool of home-based production and a high-status object of consumer desire, the sewing machine and its sojourn track the emergence and then the ascendance of the middle class as cultural ideal and social formation, along with the emergence of the female consumer and professional home manager as defining figures in Japanese modern times.

PART ONE

Singer in Japan

CHAPTER I

Meiji Machines

FIRST MACHINES

In 1841 five fishermen were caught in a storm and shipwrecked on a small island more than two hundred miles from their home in Japan. Close to starvation, they were rescued six months later by an American whaler and brought to Hawaii. The youngest of the group, age fourteen and possessed only of the given name Manjirō, stood out as curious and smart. He was befriended by the ship's captain, and in the spring of 1843 he was brought for a proper Christian education to Fairhaven, Massachusetts, next to the whaling port of New Bedford, and renamed John Manjiro. After his schooling, this extraordinary young man embarked on a three-year voyage on a whaling ship, followed by a gold-rush journey to California, before he returned to Japan in 1851.[1]

In the early years of Japan's famous "opening" to the West in the 1850s and '60s, Manjiro's mix of experience and talent allowed him to play a fascinating minor role in Japan's relations with the United States. His facility with English and his knowledge of American customs and technology won him a position as advisor-interpreter on the first official Japanese mission to the United States, undertaken in 1860 by the Tokugawa regime to ratify the trade treaty earlier negotiated by the American envoy Townsend Harris. One episode in that journey neatly represents the spirit of an era of extraordinary fascination (mixed, to be sure, with fear) about the outside world, especially the world of the "barbarians" from across the sea. Several members of the mission found their way to the laundry room of their hotel in Washington, D. C. A drawing in *Frank Leslie's Illustrated Newspaper* (figure 1) captured the "curiosity of the Japanese at witnessing the girl working one of Wheeler and Wilson's sewing machines in Willard's Hotel laundry," vividly depicting a time when even proud samurai were willing to poke around the backstage of their lodgings, and when Americans were no less curious about their visitors. The accompanying article noted that "their curiosity was greatly excited, and their inspection was close and minute into the *modus operandi* of that wonderful machine, . . . and it was understood that one of Wheeler and Wilson's sewing machines would be prominent among the most valued articles they would take back with them to Japan."[2] Manjiro was not among these men. After crossing the Pacific, his compatriots grew suspicious that Manjiro was spying for the Americans, and they left him in San Francisco when they continued east. It is not known whether any of these curious travelers actually brought home a sewing machine, but we do know that Manjiro was no less excited by this mechanical wonder; while awaiting the delegation's return to San Francisco, he bought a sewing machine to bring home to his mother.[3]

Contrary to popular belief in Japan, this was probably not the first sewing machine to enter the archipelago. Documentary evidence suggests that Townsend Harris presented a sewing machine as a gift to the shogun's wife early in 1858.[4] It is virtually certain that some of the Westerners coming to Japan in 1859 to take up residence in the treaty ports brought the machine with them as well. Woodblock prints from the early 1860s that depict treaty-port life show large-nosed Western seamstresses serving the foreign community with sewing machines (figure 2). But Manjiro's sewing machine was certainly the first to find its way into an ordinary Japanese home. (More precisely, it was the first to be turned away from a home.)

Manjiro's extraordinary story helped to establish a narrative of Japanese-American relations as a history of uplift and enlightenment featuring generous American tutors and eager Japanese tutees. It also exemplifies a process explored throughout this book: the transport, together with goods and technology, of new ways of life—and new ideas about daily life—in a world of two-way but asymmetrical exchange.

Why did Manjiro bring back this machine? At first glance the answer is simple: filial piety. But while such a sentiment was surely part of the story, a fuller and more interesting answer must recognize the modern spin he would have given to this time-honored concept. Manjiro returned from the United States in 1851 with a deep understanding of industrial

FIGURE 1. Members of the 1860 *bakufu* mission to the United States examining a sewing machine in Washington, D.C. These samurai were unlikely to have shown such interest in domestic matters back home, and this illustration conveys the fascination with which this machine was greeted by men as well as women from Japan as around the world. (*Frank Leslie's Illustrated Newspaper* [New York], June 9, 1860, p. 24)

technology; his enthusiastic explanations of steam engines, locomotives, and telegraphs had won him the respect of no less a figure than the daimyo of Satsuma.[5] He had clearly embraced an American faith in the power of machine technology. Although the sewing machine was not yet being produced for household use on a large scale during his first American sojourn through 1851, by the time of his second journey to the United States, sewing-machine manufacturers were proudly touting the product for reducing the weight of women's household labor. An 1858 flyer for the Ladies' Companion Sewing Machine "call[ed] attention of the public to these CHEAP LABOR SAVING MACHINES." The Family Sewing Machine Company in 1861 called its product "the great time-saver and benefac-

FIGURE 2. This illustration from circa 1860, one panel in a large game board that depicts various aspects of life in the treaty ports, presents the sewing machine in the early days of Japan's opening to the West as a curious object used by Western women. (Yokohama Archives of History)

tor of our race"[6] As Manjiro shopped for a machine for his mother, he surely encountered such claims and would have found them persuasive.

If so, he was moved in his gift-giving not simply by a Japanese family morality of the Edo era, but by the modern idea that material progress would uplift humanity. This was not an idea his mother would embrace. We are told she responded with disappointment, not pleasure, when she discovered that the tight stitching of the machine was poorly suited to sewing of a kimono, whose outer piece and inner lining had to be loosely stitched so the garment could be taken apart to be cleaned, and then resewn.[7]

In facing this question of the fit between the sewing machine and her accustomed dress, Manjiro's mother was among the first people in Japan to argue by her actions that Japan's material culture was an impediment to the spread of the sewing machine. Over the next century, a vigorous debate took place over the suitability of sewing machines for the stitching of what came to be called *wafuku* (lit., Japanese clothes). This neologism inscribed in Japanese culture a sharp contrast to a second new coinage, *yōfuku* (lit., overseas clothes, or Western clothes) alongside a parallel linguistic divide between Japanese styles of sewing (*wasai*) and Western ones (*yōsai*).[8]

The list of objects and practices defined in Meiji times as Japanese or Western by the first syllables of *wa* or *yō* ranged from food and drink to books and music. The incipient debate over the suitability of the sewing machine for Japanese dress focused quite specifically on the fact that one of the great merits of machine stitches—their tightness—was a demerit for a customary mode of dress that required loose stitching. This debate unfolded over the years as part of a larger discourse on the place of cultural forms and practices often newly defined as "Japanese" in a modern world understood to originate in the West. This modernity was understood at times as a threat from a West with power to colonize, at times as a resource to resist that threat, and at times as an irresistibly attractive new way of life. Contributors to the discourse of modernity in daily life, as we will see, explored the dilemma faced by Manjiro's mother as they debated the merits of what came to be called the "double" or "two-layered" (ni-jū) life combining practices marked "Japanese" with those marked "Western."

The sewing machine was therefore among those objects that carried into Japan new and at times contentious ideas concerning women's roles, the idea of progress, and the roles to be played by technology, by individuals, and by nations on the march toward an improving future. At the time of Manjiro's voyages, the idea that new technology and machines would transform both the world and individual lives for the better had been articulated clearly in American culture, but was not yet widely spread or deeply rooted in Japan. By the start of the twentieth century, the positive value of progress and the need for all individuals to contribute to the advancement of civilization and nation (or empire) was well established on both sides of the Pacific. The sewing machine would become explicitly linked to the possibility of progress for women in particular. As increasing numbers of women began to use the machine in Japan, they gave these ideas an impressive, at times surprising, range of meaning.

THE EMPRESS'S NEW CLOTHES

Until the end of the nineteenth century, actual users of the sewing machine in Japan were for the most part limited to three groups and locations: tailors and dressmakers in the handful of treaty ports where Western traders and their families had been allowed to live and work since the late 1850s; garment factories producing Western-style uniforms for the military and for the modest number of men working in modern industries, especially the railroad; and an elite world of Japanese gentlemen and ladies in Tokyo. But interest in the object and the new modes of dress that came with it extended well beyond these locales.

In the treaty ports, tailor and dressmaker shops equipped with sewing machines began to appear in the 1860s and early 1870s. Early tailors and dressmakers were Western men and women, followed by immigrant Chinese tailors from Hong Kong or Shanghai, who first appeared in 1868. Starting in Yokohama, the first Japanese tailors and dressmakers opened shops in the 1880s, almost all of them men who had learned their trade as apprentices in Chinese or Western businesses.[9] The clientele for the tailors gradually came to include elite Japanese men. Following the Meiji emperor, who first wore a Western suit in public in 1872, "the leaders of the new Japan" had adopted a style—"the top hat and black morning coat"—of recent British and North American origin that conveyed the "responsibility and self-discipline" of the rulers of a modern nation.[10] These were the men whom Basil Hall Chamberlain memorably described in 1891 as the "modern successor[s]" to the samurai, "fairly fluent in English and dressed in a serviceable suit of dittos."[11]

Female demand for the dressmakers' services increased, with a time lag of about fifteen years, as Western dress became fashionable among aristocratic Japanese women in the mid-1880s. This was the Rokumeikan era, named for a ballroom in Tokyo where elite Japanese men and women danced and mingled with the Western community. To meet the demand, Japanese-owned shops opened in the capital. Records exist for thirty-four tailor shops and just one dressmaker in Tokyo in 1880. By the decade's end, one finds more than 130 garment makers, fourteen of which served women.

These were modest numbers. Similarly, the sewing machine itself had as yet found only a limited market in Japan. Worldwide in the 1850s and 1860s, no single manufacturer stood far above others. Manjiro's machine was made by Wheeler and Wilson, one of several leading American manufacturers competing as well with German and British producers. By the 1880s, however, the Singer Sewing Machine Company had emerged as the unchallenged champion of this extraordinarily popular good. It sold eight hundred thousand machines annually, accounting for fully three-quarters of the global supply.[12]

In Japan, sewing machine sales were low, and Singer was not yet dominant. The company operated through a Belgian agent, Edward Sang, whose territory ranged from India and Ceylon to Japan.[13] Only scattered and inconsistent records of Singer's Chinese and Japanese sales activity survive from this era, but the company clearly sold only modest numbers through Sang or other agents. An 1884 report notes the sale of 635 machines over a six-month period in Yokohama (490), Nagasaki (90), and Hiogo (Kobe; 55), while an undated "Summary of Business, China and Japan, 1882–1886" lists total sales in both countries as only 117 for

1884, 355 for 1885, and 924 for 1886. The discrepancies in the 1884 reports may reflect the fact that Edward Sang was pocketing proceeds that he should have remitted to New York and covering up the related sales. Sang's malfeasance was uncovered in 1888, and Singer cut ties with him, but sales did not increase over the next decade.[14] Until the turn of the century, German producers held the largest share of the small Japanese market. Customs records from the late 1890s show the total value of sewing machine imports from Germany to be triple that of those from the United States, with British makers in third place.[15]

Compared to the modest presence of tailors or dressmakers and the limited sale of machines, the cultural impact of this object from the 1870s through the turn of the century was substantial. The history of the Japanese word for *sewing machine* offers semantic evidence for its centrality in defining what contemporaries called a new era of "civilization and enlightenment." In the earliest documents that mention it, from the 1860s, the sewing machine is named prosaically enough, either with Japanese characters that literally mean "sewing tool" or with Japanese phonetic symbols that approximated the pronunciation "sewing machine" (*shuu-ingu ma-shee-nay*).[16]

The phonetic label took hold. As it did, the adjective *sewing* dropped from colloquial use, and the word *mas-shee-nay* was distilled into just two syllables: *mishin*. In the early 1870s, this term began to appear in newspaper advertisements and articles (in 1872 with a parenthetical Japanese gloss), and the *mishin* became a sideshow attraction in Tokyo's Asakusa entertainment district.[17] About thirty years before it began to make its way into Japanese homes in relatively large numbers, the sewing machine had become an object of popular curiosity. With the name of *machine*, pure and simple, it stood as the emblem of a new era of wondrous technology.

By the early twentieth century, no less a personage than the Meiji empress was reported to desire this object. In September 1905, the *London Times* correspondent in New York described "the finest sewing machine ever made in this country . . . just completed at the works of the Singer Sewing Machine Company." That July, Alice Roosevelt, the famous twenty-year-old daughter of the president, accompanied Secretary of War William H. Taft on a large diplomatic mission to Japan. According to the *Times,* Miss Roosevelt met with the Meiji empress, who "expressed a desire to possess an American sewing machine." Informed by his daughter of this illustrious person's wish, the president placed an unusual order with Singer, asking the company to be sure that "every part of the machine where there is no friction is gold-plated. . . . A special messenger will probably carry it to Japan."[18]

No follow-up report of the actual delivery can be found in the *London Times* or other sources, but that the Meiji empress desired this good makes sense. The surprise is that she had not yet secured one. For two decades, she had been encouraging women, at least those of status and wealth, to adopt new modes of dress and new techniques of dressmaking. Both the timing of her advocacy and its ambivalent articulation reveal that Japan's rulers were acutely interested in and anxious about the daily-life impact on women of their modernization program.

The empress had moved a good bit more slowly than her husband in changing her clothes and those of the women around her. In 1872, the emperor donned Western dress for occasions of state. In 1873, the conscription law imposed military service and Western uniforms on all classes of young men. But into the 1880s the empress continued to wear Japanese clothing exclusively, even at ceremonies to inaugurate Western institutions, such as the Tokyo Normal School's kindergarten. In historian Sally Hastings's nice formulation, the imperial couple was "a mismatched pair."[19]

Women were gradually but cautiously brought into the circle of modernizing practice. In 1886 the empress began to wear Western dress in public, and the next year she issued a famous "court circular" requiring her attendants to do the same. Published in newspapers and magazines, this document had an impact beyond the aristocracy.[20] It set out model behavior not only for court ladies but for all who aspired to high social status. With a logic that may strike contemporary readers as stretched beyond the breaking point, the circular invoked ancient precedents to legitimize this change. Court regulations and changes in dress dating back to the seventh century were said to have "paved the way for today's clothing style," because Western dress resembled ancient Japanese dress, even as it was "convenient for action and movement" appropriate to the new era.[21]

With changes in dress came a call to change dressmaking. From the time of its arrival, the sewing machine had been linked to Western dress for both men and women. Like Manjiro's mother, users continued to consider the machine's tight stitching poorly suited to Japanese clothing, whether ceremonial or for daily life, but well suited to Western garments. The empress's circular echoed and reinforced this connection between modes of dress and sewing by telling women that along with new clothing, it was "only natural to adopt the Western method of sewing."[22]

At the same time, she put forth two cautionary themes that would echo for many decades. First, buy local: "In carrying out this improvement, however, be especially careful to use materials made in our own country. If we make good use of our domestic products, we will assist in the improvement of techniques of manufacture on the one hand, and will also aid the advancement of art and cause commerce and industry to flourish. Thus, the benefits of this project will extend broadly, not limited simply to the customs of dress." Second, be frugal: "In changing from the old to the new, it is very difficult to avoid wasteful expenditures, but we can certainly achieve our goal if everyone, according to their abilities, makes a special effort to lead a frugal life. These are my aspirations for the reform of women's costume."[23]

The empress and her ghostwriters were among many walking a tightrope as they advocated change for the sake of national power and progress, while seeking to protect both the domestic economy and what they identified as a virtuous tradition of frugality. Echoing the concern of Manjiro's mother, they raised issues that would emerge after World War I in a wide-ranging debate on the burden of the "two-layered life" of Western and Japanese practices. The empress and those around her were also preparing the ground for a new ideology of womanhood to take root in modernizing Japan, that of the "good wife and wise mother." One of her jobs would be to cope with this double burden.

The historians Sharon Nolte and Sally Hastings have identified the years following this imperial pronouncement, especially the span from 1890 to 1911, which neatly bridges the start of Singer's full-scale operations in Japan, as the time when "the state articulated piecemeal its official definition of women's role in industrialization."[24] They persuasively argue that the feminine virtues put forward by bureaucrats and their allies "cannot be dismissed as simply remnants of the traditional family system. . . . Their virtues included working outside the home for wages and saving in the modern banking system."[25] Two new assumptions underlay state policy toward women: "that the family was an essential building block of the national structure and that the management of the household was increasingly in women's hands."[26] The good wife and wise mother served family and society by the pursuit of education and, where necessary, employment. Fulfilling her domestic role was understood to be a public duty.

Sewing and sewing machines figured prominently in efforts to promote the ideal of a good wife and a wise mother in service of family, nation, and empire. Hata Rimuko was the first principal of the Singer Sewing

FIGURE 3. This panel from an 1893 print is one of many that depict
women using sewing machines to make Western clothing. The artist
imagines the attendants of the Meiji empress at a lakeside sewing party.
The women themselves vary in wearing Western dresses or kimono,
but they are exclusively fabricating Western wear. (Nature and Science
Museum of Tokyo University of Agriculture and Industry)

Academy, founded in 1906, and a well-known advocate of Western machine sewing. She put forward Russo-Japanese War widows as exemplars
of the "self-reliant" woman who was to be trained at her Singer Academy. Such students would learn modern skills appropriate for both full-
time household managers and those who faced the need to support their
families.[27] Hata was echoing the exhortations of the empress, who during the Sino- and Russo-Japanese Wars supported the newly founded

Japanese Red Cross, visited military field hospitals, and encouraged women to work or volunteer as nurses in Western uniform.[28]

Fascination with the sewing machine—and a sense of its multiple meanings—was also conveyed visually in this era in the popular genre of woodblock prints depicting the life of "civilization and enlightenment" as enjoyed by the symbolically powerful imperial circle. One of many examples is a fanciful 1893 print showing the Meiji empress and her ladies-in-waiting enjoying a lakeside sewing party (figure 3). This print represents a wide range of new roles for women as it vividly connects the sewing machine to various virtues and possibilities. Machine sewing, like hand sewing, could nurture traditional feminine virtues of disciplined work and care for others. But it also connected to modern transformations underway in women's dress, the activities involved in household management, and the spread of machine technology. By using this good, the print suggests, it is possible for women not only to take part but also to take pleasure in joining the nation's modernizing efforts. All these concerns would be central elements in the meanings given to the sewing machine in Japan over the following decades.

SINGER SETS UP SHOP

The Meiji empress was not alone in embracing with caution a world of imported goods and practices. Ambivalent attitudes toward the West— its power, its practices, its ideas, its people—were at the heart of Japan's modern experience from the time the first trade treaties were concluded. When the Tokugawa negotiators signed the "unequal treaties" that integrated Japan with the Western-dominated economic and political order in 1857–58, they surrendered control over setting tariffs and gave foreigners extraterritorial privileges in the treaty ports. But in exchange, they won a prohibition against foreigners doing business, owning property, and traveling or residing outside designated ports and settlements. When Japanese diplomats began to negotiate new treaties in the 1880s, the fact that such prohibitions might be abolished in exchange for regaining legal and economic sovereignty provoked fierce public opposition and near-riots.

This opposition had relaxed some when the Japanese and British agreed on a revised treaty in 1894, an agreement quickly followed by similar treaties with the other Western powers. The agreements opened the Japanese interior to what was colloquially termed "mixed residence." Anxiety certainly remained, expressed in fears that Westerners would spread evils such as materialism, gender equality, and "foreign insects poisoning the nation."[29] But with Japanese victory in the Sino-Japanese war the following year, these fears were submerged by a wave of public enthusiasm at the twin achievements of (near) equality with the Western powers and the status of empire (the Qing ceded Taiwan to Japan as a colony). In July 1899, the era of "mixed residence" began without major incident.

Although no smoking-pistol documents connect the two events, it seems certain that the opening of the interior led Singer Sewing Machine Company to end its reliance on agents and sell directly to machine users. To implement the direct sales strategy it was already using with brilliant success in most of the world, Singer had to be able to purchase or lease property for retail storefronts in towns and cities throughout Japan. The end to the "mixed residence" prohibitions made this possible. By a neat coincidence for those who like to mark historical boundaries with the Western calendar, the timing of the new treaties meant that the first steps toward "mass" marketing of the sewing machine in Japan came precisely at the start of a new century. And the fear that "mixed residence" would allow a company like Singer to spread materialism and offer a measure of power to women was not entirely mistaken.

It is impossible to pinpoint the start of Singer's direct operations to the exact day or month, but it is clear that the first Singer store in Japan opened in Tokyo sometime in summer of 1900. Yubita Kanzō, born in 1877 and the son of an owner of a sewing machine repair shop in Tokyo specializing in German machines, recalled in 1947 that, "when I was 23, Singer opened its store in Ginza and asked would I join them."[30] Three documents support this dating. A Japanese-language flyer advertising the "Singer Model 28 (Family) Sewing Machine" (somewhat oddly) states, "printed May 16, 1900, published September 1, 1900." A Japanese-language "Singer Sewing Machine Instruction Manual" for family-type models is also marked as "printed on May 16, 1900," but was "published [in] March 9, 1901.[31] And customs records for 1900, though not specific to Singer, show a fourfold increase over the previous year in the value of sewing machine imports from the United States.

We can conclude that plans under way in the spring of 1900 reached fruition with an official opening that summer. It is possible that the proximate cause of Singer's direct sales initiative was the advent of "mixed residence," while the specific summer opening in Tokyo was timed to follow the auspicious marriage of the Meiji emperor's son (later to be the Taishō Emperor) in May, 1900.[32] In any case, Singer followed by opening its "main store" in Yokohama in 1901, and it quickly expanded its network of retail outlets.

The history of the sewing machine as an item of relatively widespread purchase and use by women in thousands of homes throughout Japan— the story of the *mishin* as an item of "mass" consumption—begins with the advent of Singer. A 1903 report filed by Hata Toshiyuki, an important figure in the company's early history in Japan, teaches us something of the appeals used in the company's marketing and their connection to the values put forward by the empress in her rescript and represented in her ladies' woodblock image.

Born in 1870, Hata graduated from Tokyo Imperial University in 1899 at the relatively mature age of 29. With many of his classmates, he entered the higher civil service, in his case the Ministry of Agriculture and Commerce. In 1902 the ministry sent him for a year of foreign study in the United States. His supervisor encouraged him to visit the Singer Company and learn of its operations. Hata came away not only enthused at "the potential for this company to contribute to commerce and friendship between the United States and Japan" but sufficiently impressed to leave the ministry in October 1903 for a managerial post in Singer's Japanese operation. His enthusiasm was infectious; his wife, Rimuko, introduced above, joined him in the "family" business in 1906 as founding director of the Singer Sewing Academy.

In the report he submitted to the ministry, Hata nicely conveyed the sense of mission of C. H. Pierce, head of the Singer Export Company. Pierce was not willing to accept the idea, certainly bad for business in a place where Western dress, especially for women, remained uncommon, that sewing machines were incompatible with the fabrication of Japanese dress: "Singer's object in selling machines in Japan is not limited to Western sewing, since Singer machines can also be used to sew Japanese clothing."

Pierce hoped to challenge established gender norms as well. Hata remarked that Pierce made the case "with passion and conviction" that "the situation of Japanese women is pitiful, for all the housework falls on their shoulders, as does the burden of serving parents and caring for children, so that the burden on Japanese women for the sake of the progress of the nation is too great to bear. If Japanese women used Singer machines, they could do in an hour the work that normally took a day. So, selling Singer machines in Japan will not just serve the cause of corporate profits, but will contribute to the advance of Japan's material civilization."[33] Armed with such a vision, which it put forward in other non-Western and colonial contexts as well as in the West, Singer was well positioned to sell its sewing machine as a necessity for the "good wives and wise mothers" whose mission was to work with their husbands and families to build Japan as a modern nation and empire, a place of wealth and power.[34] The company was both reflecting and helping to generate new ideas about good wives and wise mothers in the years when this role was being defined and articulated by state officials, including the empress. As Singer established itself as the premier seller of sewing machines in the land, it directly and indirectly shaped the practice and understanding of modern life and the consuming household. And in some measure, it allowed women to push at the boundaries of their prescribed new roles.

WHO SEWED FOR WHOM, AND WHERE?

However limited, the initial spread of the sewing machine in Japan and the early steps in the transformation of sewing and dress practices were part of a global story. The sewing machine spread rapidly around the world in these same decades, and its diffusion was the product principally of a single company with a remarkably effective and uniform selling system. Everywhere it quickly moved from the realm of marvel and novelty to occupy an important place in the fabrication of garments in factories, small workshops, and homes. But if the story was in large measure global, it is nonetheless true that questions of who sewed what, for whom, and where were answered differently from place to place. To understand modern patterns of use and negotiations of meaning in Japan, we must briefly examine the preindustrial organization of garment production and consumption in a comparative frame.

It seems fair, if sweeping, to generalize that prior to the nineteenth century, the majority of sewing and fabrication of dress around the world took place in the home. In Europe, as Anne Hollander notes in her study of modernity, gender, and dress, "It must always be borne in mind that until the last third of the nineteenth century . . . there was virtually no ready-to-wear fashion for women, only outer garments and headwear. What was not custom-made was homemade or second hand. In fact, most women, rich or poor, knew how to sew or understood sewing."[35] This statement applies as well to non-Western societies, where there was an even greater proportion of home-made than custom-made wear.

The phrase "what was not custom-made" in Hollander's generalization points to the critical variable factor, especially as we consider the appearance of the sewing machine in Japan along with Western sewing and dress. Before the sewing machine spread globally, one finds differences around the world in the prominence of the specialized professions of tailors or dressmakers who custom-made garments for multiple customers, as well as paid seamstresses who stitched together particular clothing pieces either at home or in workshops. It appears that in the eighteenth and nineteenth centuries, a divergence took place that set much of Europe, along with its American colonies and new nations, apart from other places, Japan among them.

In the former, before and during the industrial revolution but well ahead of the spread of the sewing machine, the fabrication of clothing first for men and then for women moved out of the home and into the hands of specialized producers: tailors and dressmakers as well as pieceworkers. Hollander notes simply that "huge numbers made their living" by this artisanal handiwork.[36] In Britain, by the early nineteenth century production of clothing outside the home, whether by artisans or in garment factories, had become widespread.[37] In her study of "the female economy" of dressmakers and milliners in the United States, Wendy Gamber dates the takeoff of these trades from about 1800 through the 1840s, sparked by what she calls an "antebellum consumer revolution." By 1870, dressmakers and milliners comprised the fourth largest women's occupational category in the country, and—of importance in thinking comparatively—they "did not simply transfer domestic skills to the marketplace; more often than not, they learned their trades in the workshop, not the home."[38] The subsequent story of garment production in the United States, in Gamber's telling, remained outside the home, as a struggle unfolded between garment shops, where independent dressmakers made clothes to order, and factory or workshop production of ready-to-wear clothes sold in department stores. By the 1920s "the factory claimed an undisputed victory over the garment shop."[39]

In Japan, in sharp contrast, through the end of the Tokugawa era (1868) most clothing was fabricated in the home.[40] Liza Dalby nicely summarizes the situation at the moment when Western dress (and Westerners) appeared in the treaty ports in the 1860s. "Traditional dress had not required tailoring. Dry goods merchants sold bolts of cloth in a standard kimono width, and women of the household made it into the required article of clothing. Sewing at home formed a major part of women's work, for although it was not difficult—straight seams and minimal piecing—it was constant." The time demand resulted in significant measure from the fact that traditional dress "was taken apart and resewn at each washing."[41] The scientific time survey had yet to be invented, of course, so comparison with later practice or other places is not possible, but on the eve of Japan's modern revolution, sewing was clearly a time-consuming element of daily life.

Outside the home, used clothing was sold to relatively impoverished city dwellers, but there was no significant commercial market for ready-made or custom-made new clothing.[42] Dalby's concluding point is important as we trace the continued importance of home-based sewing and the time it required, whether by hand or machine, whether for Japanese or Western dress from the 1860s forward: "Since the making of clothing was not a professional activity, there was no pre-existing group that might logically have made the switch to sewing yōfuku. The profession of tailor [and we can add, that of dressmaker] had to be created from whole cloth."[43]

As Western dress began to spread during the 1870s, especially among men, and as sewing machines entered Japan as well, the organization of garment production changed significantly. Not surprisingly, the direction of change reflected these initial conditions, even as an important new gender divide emerged in the production and consumption of clothing. We have seen that men in the offices of the new government and in modern businesses turned to Western suits as early as the 1870s and '80s, and Western wear was adopted early on for military uniforms and increasingly for male factory laborers. By the late 1880s, Western pants (zubon) had completely displaced traditional men's wear such as leggings (momo-hiki) or the skirtlike hakama.[44] Women in the home neither possessed nor acquired the tools or the skills to make Western menswear, so that by the end of the nineteenth century, the new profession of the tailor, wholly the province of men working outside the home, had become well established, along with the production in small workshops, again staffed by men, of ready-to-wear Western clothes for lower-level clerks and the like.[45]

The story for women, both in the production of dress and their preferred styles of wear, was quite different. A few women set themselves up in dressmaking businesses outside the home, but their numbers—fourteen in Tokyo by 1890—hardly compared to the growing corps of tailors—well over one hundred—who were fabricating menswear.[46] This vanguard of dressmakers was a noteworthy portent, but a tiny number in a nation of close to twenty million women in 1890. Western dress for ordinary women would gradually spread, first to young girls, students, and young "working women," from the early 1900s into the 1920s, and then in the 1930s to a growing minority of adult women. But customary forms of dress—now dubbed *wafuku,* or "Japanese clothing"—remained the preferred choice for the majority of adult women. And not only this *wafuku,* as in the past, but also those forms of Western dress that did win some adherents, were still sewn in the home by women for women.

It was thus most common by the late nineteenth century and into the twentieth to find male tailors in business outside the home serving male customers who wore these clothes to work, while women inside the home produced, laundered, and repaired both Japanese- and in some cases Western-style dress for themselves and for their children. This remained a time-consuming element in their daily routine or, for the wealthy, the routine of their servants. It is tempting to map such a gender-divided system of production and consumption of dress onto a divide between a modernized world of men's wear, professionalized and commercial, mechanized and scientific, attuned to national and global trends in fashion, and a traditional realm of women's dress, the domain of amateur home sewers. But such a bifurcated perspective is wholly misleading. Rather, although the location of sewing in modernizing Japan—outside the home for men, inside for women—was gender-divided, the form of its practice was not. As it did for the tailor shop, the sewing machine and new modes of sewing and garment production would bring modern life directly into the home in the form of professionalized training, a new science of home economics, and market-oriented, mechanized fabrication of clothes. It also helped create the role of the women as consumers-in-chief of the modern home.

CHAPTER 2

The American Way of Selling

When the Singer Sewing Machine Company entered the Japanese market in earnest in 1900, it brought a half-century of experience as "the world's first successful multinational company."[1] Founded in 1851 by Isaac Merritt Singer, the company focused on world markets and household users from the outset. By 1864, exports accounted for 40 percent of the machines it sold. By 1880, Singer claimed almost half of the global sewing machine market. Its worldwide sales that year surpassed five hundred thousand units. By the eve of World War I, shipping its product from factories in the United States (New Jersey), Scotland, and Russia, Singer controlled 60 percent of the American market in household machines and probably 90 percent of foreign markets.[2]

Business historians have naturally been intrigued by this remarkable company. In explaining its success they have reached a persuasive consensus. The company's unchallenged global position derived not from low cost; its machines were relatively expensive. Nor was it grounded in technological superiority; other American and German machines were reputedly better. The secret was neither a more appealing design than others, for all the leading machines looked alike, nor was it more effective advertising. The Singer magic lay in "its direct selling organization knocking on people's doors all around the world."[3]

Singer was founded and headquartered in the United States, where its forty-one story building in lower Manhattan was the tallest in the world when it opened in 1908.[4] Developed in America, the renowned selling system which it brought to Japan in 1900 had by then taken firm shape as a product of trans-Atlantic and transnational experience. The company drew on selling experience gained in Britain to devise its famous canvassing system. A large number of German managers were sent to run Singer's early business in Russia. Against this background, it is no surprise that in early twentieth-century Japan Singer proudly advertised the sewing machine as a good in the vanguard of global civilization, embraced by "Europeans and Americans . . . of all classes high and low."[5]

As Singer set up its operation in Japan, it made remarkably little adjustment to its selling system. In less than a decade, Singer's shops had spread throughout the country, and Singer used its standard global model in Japan with fair success into the early 1930s. Annual sales of its household machines reached twenty thousand in 1911, and roughly fifty thousand per year by the early 1920s. By the 1910s, Singer stood unchallenged with more than 80 percent of the Japanese household market.[6] As elsewhere, it had consolidated a position so strong that its brand name came to stand in—like Hoover—for the product itself. By the end of the 1920s, Singer employed about eight thousand people. These ranged from salesmen who worked on commission to young women instructors and salaried staff in the head offices and warehouses. Singer's management was quite fully localized; no more than a dozen American, Canadian, or European men worked as expatriate executives at any one time. Measured in employees, Singer was the largest foreign corporate presence in Japan.

Although desire far outran the ability to buy it, the Singer sewing machine by the 1920s had become one of the most wanted possessions for that significant minority of Japanese households, mainly urban, who defined themselves as part of the nation's new middle class; it was a practical and desirable possession as well for women who were part of the working class. Looking at the decades from the turn of the century through the early 1930s, this and the following chapter examine Singer's efforts to sell, the American way, and its role in selling the American way. Singer implemented virtually unchanged a system of retail selling that, although created through trans-Atlantic trial and error, came to be understood in Japan as fundamentally American. The Singer system helped produce new social roles and forms of discipline— on the supply side for men as sellers and women as teachers, and on the demand side for wives and husbands as installment plan borrowers. Singer at the same time fostered a way of life, for women in particular as home managers and consumers, that was identified with America and modernity, even as the company and its product provoked a belief that daily life was a problematic, hybrid undertaking that retained "Japanese" characteristics.

THE SINGER SYSTEM IN JAPAN

As it did around the world, Singer set up a multitiered organizational pyramid in Japan. A head office, in this case for the "empire of Japan," oversaw a small number of "central" offices, which in turn coordinated a network of regional or "branch" offices. Five or six regional offices stood under each central operation and supervised the many retail stores, also called local offices, that sold directly to customers.[7] The central and regional offices were staffed by salaried employees who managed accounts and inventory, distributed the products and parts, and oversaw the retail shops. These local shops employed a manager and his staff of salesmen, a separate group of installment collectors, young women instructors, repairmen, and transport workers. The managers, sellers, and collectors worked on straight commission. The instructors were paid a weekly wage, but they were also expected—as a condition of continued employment—to drum up sales, for which they received a commission. A 1922 photograph of employees in front of a Singer shop illustrates this division of labor (figure 4).

In his 1902 report to the Ministry of Agriculture and Commerce, Hata Toshiyuki extolled the cost efficiency of the Singer organization. Singer, he claimed, was in the forefront of American manufacturers who had pioneered ways to sell directly to the public. Anticipating arguments that have been made for at least a century since that time, he explained that if large-scale producers in Japan could cut through layers of distributors and retailers and sell directly, as Singer did, "they could provide the newest inventions to all people at the lowest prices."[8]

Hata was correct to conclude that Singer had devised an effective sales system, but he was wrong to call it cost-efficient. Even though most of the employees posing for the photograph in front of the retail store in Mito (figure 4) were paid only when they sold a machine or collected a payment, the sales network was expensive. The numerous leased stores and the large salaried staff in the central and regional offices added up to a substantial fixed cost. Andrew Godley notes that in the United States, "flooding the country with shops, canvassers, and collectors was a very expensive method of selling."[9] His comparative and global analysis of the Singer selling system concludes that its selling organization tended to "maximize sales rather than profits."[10] In Japan as elsewhere, it dom-

FIGURE 4. This photograph, taken in 1922 in front of Singer's Mito store, about a two-hour train ride north of Tokyo, presents the full range of employees at a relatively well staffed shop. From left to right we see the shop's transport worker, three young women teachers, and thirteen men who are either sellers or installment collectors. The man standing precisely in the middle of this array, wearing a hat, is very likely the shop manager. Turnover was high among the sellers. The visible portion of the sign behind the second salesman from the right reads "Salesmen," and the obstructed portion very likely continues with "Wanted." The second story housed the shop's sewing school. (Mochizuki Yoshimasa)

inated the market not because of low cost but despite relatively high cost. It did so by building an omnipresent network not only for selling its machines, but for instructing buyers in their use and offering reliable maintenance and repair.[11]

Probably because of proximity to the docks and customs house, Singer's Japanese corporate headquarters started out in Yokohama; it was moved to the port of Kobe in 1924 following the devastating Kantō earthquake of 1923, only to move back to Yokohama in 1933 after a bitter labor dispute. By the 1920s, Singer maintained central offices in Yokohama, Kobe, Osaka, and Seoul. The Kobe office oversaw all regions to its west (including Taiwan); the Osaka operation superintended the regions of central Japan; and the Yokohama central office coordinated regional branches from Kantō to the north. The Seoul central office oversaw a growing operation in Korea and Manchuria.

After opening the first store in Tokyo in summer of 1900, Singer built this network at an impressive pace. Actual lists of stores are fragmentary, but the trajectory is clear. Hata's report claimed that Singer had by

1903 opened stores in eight cities: Yokohama, Tokyo (where there were most likely several stores already), Osaka, Kobe, Nagoya, Nagasaki, Shizuoka, and Maebashi.[12] A typewritten global "Directory of Shops Under Controlling Agency," undated but most likely from 1904, lists fifty-one stores in Japan, and a 1906 directory reveals that Singer's Japanese network boasted seventy-one retail stores from Sapporo in the north to Kagoshima in the south (one per city, except Tokyo, which had eight). Japan at this time included the colony of Taiwan, where Singer had opened a single store. Korea, in 1906 a Japanese protectorate but not formally a colony, was listed separately, with three stores.[13] Sales figures follow a similar upward trend, from 6,529 machines sold in 1903, the first year for which reliable data are available, to 12,895 in 1906 and more than 20,000 in 1911. By the end of the 1920s, the company boasted a Japanese operation of roughly eight hundred stores and eight thousand employees.

At the heart of the Singer system stood the salesman, or canvasser. A retrospective account from 1951, most likely the testimony of a veteran Singer salesman, concisely summarizes the canvasser's daily routine in the prewar era, as well as the strategies and support personnel that made him effective:

> A sales office was set up, with a goal, say of 100 machines in monthly sales, and there are three levels of employee. An office chief, said to be a veteran with strong powers of observation, salesmen, and assistants. The former go door-to-door in a given district in the morning, and record a set array of information onto customer prospect cards. These are turned into the office at noon each day. The office chief then selects one or two good prospects. The salesmen are then sent to make second visits to these. When a few strong prospects are thus identified, the salesman and a teacher make a visit, with a machine in hand, and give a demonstration. Then they leave the machine for a free trial period, during which customers can get free lessons. By now, a sale is understood to be assured.[14]

This account touches on several key aspects of the Singer way of selling. These were not spot transactions; the salesmen were building long-term relationships. Success required keen understanding of the social world of their clientele, including an understanding among male sellers of the female customers' lives and desires that was not easy to come by. Yamamoto Tōsaku, a Singer salesman who would in 1932–33 lead the dispute of Singer employees, recalled in a 1948 discussion that most sales during his time with the company (1917–1933) were to well-to-do families: "First we would look into the entryway to see what sort of footwear there was, and maybe the daughter of the house is 3 or 4 years old, so we'd plan to come back in several years to try to make a sale." Yamamoto griped that "selling Singers was really hard. The buyers were women so it was particularly tough. At the time, they had no knowledge of machines, and it was the era of hand sewing, so it was really tough to sell them a Singer sewing machine."[15] Another discussant, Sekine Harukichi, recalled working one home for eight or even ten years. Yamamoto added that "to sell even one machine, you needed to butter up the cat, too. The cat wasn't stupid! If you kicked the cat, the deal was off!"[16]

For those who already owned sewing machines Singer's aggressive trade-in policy, also part of its global system, was crucial. A late nineteenth-century American ad emphasized Singer's willingness to take in old machines, boasting that "such [imperfect competitor] machines ought to be put out of existence. We therefore offer to exchange our new and latest improved Machines for old Sewing Machines of every kind, on liberal terms. Such old Machines as are thus obtained by us will be destroyed."[17] In Japan, this practice, together with an abundant inventory of parts, allowed the company to literally smash its competition, in particular the German makers who dominated the Japanese market before Singer began full-scale operations. Singer's early leaflets in Japan noted prominently its willingness to take old machines of any make for trade-in, and Yamamoto and other Singer veterans recalled taking in any and all German machines on trade-in and then simply destroying them to eliminate any source of spare parts or other secondary market for the competition.[18] Only a company with tremendous capital resources could afford this costly competitive strategy, since it could have resold the machines.

The free trial noted in the 1947 roundtable and featured prominently in Singer advertising leaflets was another crucial entry tactic. Like the abundant parts inventory and the wasting of trade-ins, the free trials made the Singer selling system expensive. A second retrospective account estimated, less optimistically but more believably than the 1951 recollection quoted above, that 20–30 percent of those who took in machines on trial actually purchased them.[19] This was no just-in-time inventory system; to offer all these trials, Singer had to keep its retail stores stocked with an inventory much larger than its monthly total of final sales.

Over time, the household machine and home-based users, more than garment factories, tailors, or dressmakers with storefronts, came to constitute the great majority of Singer's market in Japan. Selling to these homes was a team effort that aimed to create demand for a product whose use was not yet well known. Male sellers and female teachers were themselves being trained in—and at times chafed at—the modern science of selling. At the same time, they were educating their customers in the ways of a new world of getting and spending.

THE MAKING OF THE JAPANESE SELLING CLASS

Selling is intrinsic to capitalism. The statement is obvious, yet the social and cultural history of salespeople is far less studied than that of the workers who made the goods. One exception is a recent work by Walter Friedman entitled *The Birth of a Salesman* that deals with salesmanship in the United States.[20] Friedman shows that the sellers of the Singer machine played a key role in bringing system and reputability to a job whose earliest practitioners were notorious for lacking both. He locates the transformation in the 1880s through the early 1900s, when a shift took place from the strategies and culture of the "agricultural canvasser and the wholesale drummer" to "a newer, more aggressive, and highly managed form of salesmanship." At the forefront of this transformation were mass manufacturers who "began to build their own cadres of salesmen, and in doing so, developed the first modern sales forces, . . . set[ting] the pattern for companies in the decades that followed."[21]

Central to the endeavors of Singer and other large corporations who gave birth to the modern American salesman was the application of a standardized system to what came to be considered the science of selling in a process parallel to the more famous revolution in "scientific management" of industrial production pioneered by Frederick W. Taylor.[22] Friedman begins his analysis of "scientific sales management" in America with Singer and the sewing machine: "The sale of the sewing machine paved the way for other expensive or intricate machines," from the vacuum cleaner to the automobile, whose makers in the early twentieth century also relied on door-to-door sellers who provided information, education, and credit as they struggled to convince customers that the high prices of their goods was justified.[23]

In Japan as well, Singer was a leader in the "science" of systematic selling. The company trained salesmen in the protocols of comportment and dress, and taught lessons in communication and psychology. But with Singer among the midwives, the birth of the salesman in Japan—not unlike the making of its working class—was marked by considerable tension and resistance to the imposition of managerial discipline. Among the salesmen in the belly of the beast of modern capitalism—prominently including the Singer sales force—one finds important ambivalence in response to the logic of the market, especially to elements understood as an American way of doing business.

The terms of the Singer salesman's trade in Japan, as elsewhere, were stringent. A canvasser earned a 12 percent commission on each machine sold. For cash sales, he received the entire commission at once, but for credit sales, he received a 10 percent commission when a sales contract was signed and a down payment taken. He was paid the remaining 2 percent over three months, pending the successful collection of the monthly installments. The Singer collectors earned a 7 percent commission, likewise pegged to their successful harvest from customers. Few records survive that document individual canvassers' income in Japan, but a rough calculation from the aggregate figures makes it clear that many of the sellers and collectors must have taken home thin pay envelopes. In the late 1920s, a time when the company employed between seven and eight thousand people, including salaried office staff and teachers, as well as canvassers, collectors, and branch managers paid by result, annual expenses stood at $3 million per year maximum, roughly $450 per employee. Even if the company's local expenses had consisted almost entirely of wages, at the exchange rate of two yen per dollar, the average income would have been less than eighty yen per month, roughly the entry-level pay of a male white-collar worker at the time. But of course the "expenses" category—unfortunately not broken down—included rent, transportation and warehouse costs, advertising, and utilities, so the net average income of Singer employees was much less than the earnings of the typical salaried office employee of the era. The 1932–33 dispute provoked one employee to enumerate in detail his income and expenses in a leaflet. His particular figures roughly square with this aggregate calculation if we assume that about half of the overall operating expenses went to wages: this Tokyo-based payment collector working on commission took home a net monthly income of 46 yen.[24]

Given these figures, it is no surprise that turnover was high. The world of the sellers included two groups: a small group who thrived and excelled, such as the passionate Yamamoto Tōsaku, who rose within two years to become manager of the most successful local store in the nation; and a large group who entered and exited quickly.[25] In the photo of the Mito shop (figure 4) a signboard proclaimed "salesmen wanted," visual evidence of the chronic need for new troops to carry the Singer message door to door. The protesting employees in 1932 argued that punitive terms of employment caused high turnover, offering as evidence a Yoko-hama store that had to hire thirty-one different salesmen in one year to maintain a staff of nine canvassers.[26]

For these sellers, discontent went beyond low pay. Even the success-ful men were frustrated at the uncertainty of their employment and what they considered unfair disciplinary practices; the high-flying Yamamoto, after all, led the Singer dispute. The mandatory "fidelity insurance plan," used globally by Singer to protect the company from losses, generated strong criticism. It charged Singer canvassers and collectors annual pre-miums of ¥5–¥10 and covered from ¥500 to ¥1,000 of losses if, for ex-ample, a customer absconded with machines before completing install-ment payments. Singer added additional layers of protection by making the sellers responsible through pay deductions for losses generated by customers; in addition, each seller had to deposit ¥200 with the com-pany, from which penalty charges could be drawn, and each was required to list two guarantors who would also be responsible for any losses.

Employees argued that the company profited hugely from this multi-layered system of protection; they questioned whether the fidelity insur-ance premiums were actually turned over to an insurance company and asked that insurance certificates be issued to each employee. The sales-men also complained bitterly and repeatedly over the assessments levied on sellers, canvassers, or store managers as pay deductions. Not only when a customer absconded with a machine not paid for, but when he or she terminated an installment contract prematurely on the one hand, or paid off the principal balance ahead of time at a cash discount on the other, Singer deducted from the sellers' and collectors' wages the portion of the commission that, by Singer's logic, had been prepaid in anticipation of installment revenues now lost.[27]

Discontent with these policies led Singer employees to launch two small dispute actions in the mid-1920s, one in Tokyo in March 1925 and an-other led by employees in Osaka area stores from December 1925 through January 1926, as well as a major struggle from September 1932 through January 1933.[28] By the time of the 1932–33 dispute, the employee as-sociation was bitter enough to characterize itself as heir to a "thirty-year history" of enmity with the company. It claimed that nearly thirty dis-putes, large and small, had taken place over conditions of employment since the company began operating in Japan.[29]

As these discontents simmered and sometimes boiled over, Singer's top managers in Japan, who in the 1920s included several Americans, one British man, and one Canadian, sought to impose more effective discipline by insisting that the treatment of employees was reasonable, that the seller's vocation both served society and offered realistic reward for hard work, and that in any case, the globally tried and true Singer system could not be changed. A lengthy "Statement" given by Singer to the Japanese Foreign Ministry in November 1932 forcefully made the case that "our business in Japan, in addition to having given profitable employment to many thousands of men and women," has made "machine sewing an essential part of a girl's educational training," enabling girls to "form new homes" and sustaining the "future motherhood of the nation."[30] As the New Year holiday came to an end in early January 1933, E. L. Vest, director of the Osaka Central Office, sent a letter to the two hundred store managers in his jurisdiction offering a glimpse of his definition of scientific selling. The economic outlook had improved, he claimed; many families long desiring to buy sewing machines for the sake of their children's education would be able to do so. To reach these customers, "energetic door-to-door selling is indispensable," and "there is no better method than setting out orderly blocks [of sales territory] from early in the new year . . . . The salesmen must of course cover their blocks every day systematically and in proper order, and as we all know, one day's sales bring more the next day."[31]

The many "how to" books published in the interwar era aimed at salesmen (and some saleswomen) show that Singer was typical in thus invoking system and social progress to cultivate a proper selling mentality and behavior. One early work of this genre, dated 1916, was entitled *Scientific Business Strategies: Salesmen and Selling Tactics,* a name that evokes Taylor's *Principles of Scientific Management,* translated three years earlier into Japanese.[32] There is little to distinguish these books from the persuasive and educational tracts aimed at sellers in the United States, and they freely drew on American sources and anecdotes. Their advice echoed Singer's position that there was no need to adapt the system of selling in Japan to the particularities of the local scene. They typically distinguished between the older, unsavory and unsystematic selling practices of the itinerant peddler and the newer, honest and respectable scientific endeavors to advance a project of civilization and progress. *Scientific Business Strategies,* for example, began with a chapter on "the mission of the salesman," who is "the publicist for the happy sound of civilization."[33] He convinces people to "replace carts with bicycles, pencils with typewriters, (oil) lamps with electric lights, and needles with sewing machines." None of these modern goods, in this author's telling, were of obvious value or utility to the "extremely ignorant" ordinary folk of the world; without the persuasive power of the salesman, society would not march to the tune of progress. Twenty years later, Kuramoto Chōji echoed this theme, claiming that without the efforts of salesmen, "who create the desire for civilization," early users (in the United States) had been unwilling to try goods such as sewing machines, whose utility seems so obvious now.[34]

Reading through these tracts, one senses in the repetition a frustration that core messages were not being heeded. Ishikawa Rokurō in 1925 argued that although people generally scorned the salesman, this professional seller's work deserved respect.[35] Shimizu Masami, the most prolific producer of "how to sell" tracts, similarly claimed in 1925 that "the old peddler has no trust or respect; . . . the new salesman has trust as his cornerstone." An expanded version of this book in 1937 essentially repeated his case.[36] In prewar Japan, it was an uphill battle to convince both sellers and their customers that theirs was an honorable profession on a par with the work of salaried managers in the offices of corporations or state bureaucracies. Only in later decades, along with their efforts to systematize employment relations for office workers and factory laborers, did many companies effectively cultivate esprit and discipline among sellers, who more readily than before (and never completely) accepted the terms of their employment as natural and respectable.

THE SELF-RELIANT "WOMAN TEACHER"

A small cadre of working women joined male canvassers and collectors to educate users and create demand for this unfamiliar product. They were called, simply enough, "women teachers" (onna kyōshi). If the salesman stands as metonym for the salaryman (sarariiman), a male emblem of the new middle class of the early twentieth century, these women also stand in for a larger category: the "working woman" (shokugyō fujin), who would emerge to prominence in the interwar era. The term women teachers in present-day Japanese usage has a pejorative ring, but in their day these women were stylish symbols of independence and modernity; a petition by Tokyo-based teachers during the 1932 labor dispute complained that the company's treatment betrayed the "beautiful title" of "woman teacher."[37] Three of them can be seen standing proudly in front of the Mito shop in 1922 (see figure 4), wearing what at the time was the latest fashion for women, the hakama, previously a garment for men. Often trained at the Singer Sewing Academy in Tokyo, they would offer instruction for a fee at smaller Singer sewing schools located on the second story of retail stores around the country. In addition, they visited buyers to offer free home lessons, their number pegged to the cost of the machine.

These women shared with their male coworkers mobility and the privilege—to an even greater extent than the salesmen or collectors—of entering the homes of strangers, a practice that had little if any grounding in female roles of earlier times. They were also the focus of a discourse on self-reliance for women, likewise not found in the Tokugawa or early Meiji eras. The prominence of this ideal confounds a retrospective assumption that middle- or upper-class Japanese women of this era were expected to remain economically dependent on fathers or husbands; rather, women by the early twentieth century were expected—where necessary—to be self-reliant.

Hata Rimuko, in addition to serving as the first principal of the Singer Women's Sewing Academy, wrote a text for her students in 1908, *Learning Machine Sewing By Yourself,* which became a bestseller on the open market as well. In a later edition, she described her initial intent as "to provide an appropriate occupation to the hugely increased numbers of war widows and survivors of the Russo-Japanese War."[38] While her marriage to Hata Toshiyuki—by this time the head of Singer's main office in Tokyo—surely helped her get this job, nepotism was not the main story. Before marrying Toshiyuki, Rimuko had graduated from Tokyo Normal School, the nation's premier teacher-training institution, and before taking the position at Singer, she had taught in three different private and public girls' schools, taking breaks for childbirth or child rearing.[39] She was an experienced teacher, 31 years old, when she took on this new position.

The school's mission statement presented economic independence as one among several objectives: "This academy will teach necessary academic skills and morals to housewives, to women wishing to be housewives, and to women engaged in charitable works, or women who have need for supporting themselves independently." Self-support here stands at one end of a continuum from more purely domestic roles through a wider civic realm of presumably upper-class philanthropy. The statement went on to place the school's mission, and women who could sew, at the vanguard of a larger modernizing project: "The sewing machine saves on household costs and allows economical use of women's time and labor. Further, [the school] aims to teach how to make clothing in accord with progress of the times."[40]

The school offered multiple curricular tracks, some vocational and others aimed at housewives interested in sewing for their families. Enrollment records are scarce, but a re-accreditation report filed in 1919 notes that the school enrolled three hundred students, about half of whom lived in the dormitory.[41] Many of its graduates went on to take jobs as Singer instructors. In addition, by the 1920s a growing number of sewing schools offering both vocational and homemaker tracks had opened for business; they too provided training and jobs for "women teachers."

The key word in the Singer school's appeal to economic independence was *self-reliance* (*jikatsu*), a virtue prescribed with remarkable frequency to Japanese women in the first several decades of the century. It appeared in the headline of an advertisement aimed at prospective students of the recently opened Singer Sewing Machine Women's Academy: "A Call to Women Seeking the Road to Self-Reliance" (figure 5). The ad's text notes that the school was overwhelmed with employment requests for its prospective graduates for jobs as sewing teachers in homes or in schools. It was unable to meet this demand with current graduates. All students who graduate are promised jobs.[42]

One woman who heeded this call, and who perfectly fit the profile sketched in the school's mission statement, was Sugino Riu. Her modest fame in later years surely reinforced the persuasive power of such ads. Sugino was the wife of a distinguished naval officer, Sugino Magoshichi (b. 1866), who raised himself from poverty to become a low-ranking officer in the Sino-Japanese War and then lost his life in the Russo-Japanese War. Riu received a grant beyond the ordinary pension in recognition of what the Navy saw as Magoshichi's heroic service, and she used a portion of the award to erect a statue in his honor. The rest she reportedly used to raise and educate three children. She was one of the first students of the Singer Sewing Machine Women's Academy. Upon graduating in 1907, she took a job as sewing teacher at a higher girls' school in Mie prefecture, her late husband's home. It is difficult to pin down Sugino's mixture of reasons for pursuing her career. Given that she could afford to spend a portion of her state reward on the statue honoring her husband, it seems unlikely that she was in desperate straits; her decision to become a sewing teacher very likely reflected a desire to play a meaningful role in society beyond meeting economic need.[43]

A second model student in the Singer Academy's first class came all the way from the southernmost island of Kyushu. Hirooka Kiyo was already employed in the local schools as a young teacher of Japanese-style sewing. Hata, it is clear, recruited actively and widely; through a veteran sewing teacher of her acquaintance, she learned of Hirooka and en-

FIGURE 5. This November 1906 advertisement in *Fujo shinbun* (The women's newspaper) is addressed to "Women Seeking the Road to Self-Reliance." It depicts the school's building in central Tokyo, an impressive three-story structure, and offers details of curricular choices and costs. The least expensive "basic track" was one yen per month, about one day's pay for an entry-level (male) job in a company or government office.

couraged her to come to Tokyo to study Western sewing. Within a year of opening the Tokyo school, Singer set up a school in Kobe to serve the Kansai region and hired Hirooka for a post there. She went on to help found the Singer school that opened in Nagasaki (on her home island) in 1909. After two years, she returned to her hometown and married.

Having taken her husband's family name, she opened her own school as Ochimizu Sewing Academy. She exemplified Hata's success in giving women skills to support themselves. She in turn inscribed Hata's goal in her own school's mission statement: to offer housewives the knowledge and skills to open their own businesses.

To what extent, if any, did "women teachers" with their earnings and their mobility—the Singer employees offering lessons in homes or company schools or the teachers in other private sewing academies—challenge expectations of the proper role for women? Jordan Sand describes a 1920 essay in the monthly magazine *Shufu no tomo* (The housewife's companion) written under the given-name byline of "Tsuruko." It ran as part of a series in which staff writers visited new homes to describe the latest trends in architecture and modern living. For Sand, whose view I find persuasive, Tsuruko's mobility, her description of the home in romantic terms, and her boldness in approaching and entering a stranger's residence, were "an unusual stance for a 'housewife' of the time, pursuing and visually possessing the object of her desire."[44] I see the "woman teacher" who appeared under Singer's auspices fifteen years earlier to be cut from a similar mold. To be sure, neither the object of Tsuruko's desire (as Sand takes care to note) nor the subject taught by the Singer instructors stands outside or opposed to the domestic realm that women of the time typically inhabited. Even so, the mobility and earnings of such women held the potential to challenge male prerogative and power, or at least the male sense of such power.

In the mass print media one does find traces of anxiety at young women stepping into society in this fashion. A 1932 story in a popular women's magazine extols an ambitious young Japanese woman living in Taiwan who founded a successful clothing business. She had started her career in 1929 as one of Singer's "woman teachers" but shifted to home-based production and sale of clothes when her mother objected to her work outside the home.[45] A far more spectacular concern was summed up in a newspaper headline just four months later: " 'The *Mishin*'s Clatter.' Actually, Red Propaganda. *Mishin* Teachers Arrested." Four women sewing teachers had been arrested on suspicion of using their work as cover to revive a union of garment workers recently banned for alleged communist ties.[46] Looking at women teachers of all sorts, a conservative critic in 1935 penned a three-part critique of the "Woman Teacher Problem" lamenting that teachers would learn to love their wages more than their children. Materialism would invade the home, and Japan's family system would collapse.[47]

Such overwrought perceptions of a threat from the mobile and self-reliant sewing (or other) teacher, while important to note, were relatively isolated. The woman teacher was more often praised or envied than feared or scorned. The most important reason for the relatively calm response to this figure likely rests in the key phrase *just in case*. Self-reliance, that is, was typically praised as a virtue of importance "just in case" a male source of support died or for other reasons failed in his obligations. In the early decades of the twentieth century, any fears that the female teacher and others like her would break the bounds of this conventional domestic life were outweighed by the desire to train women to be active and potentially self-reliant supporters of family, nation, and empire.

THE PROMISE OF CONSUMER CREDIT

The home visits of these teachers were connected to what Singer's Japanese managers justifiably saw as the most important innovation in Singer's entire arsenal of sales practices: the offer of installment credit. The weekly or monthly "free" lessons provided a strong incentive to continue paying the installments to the collector, whose monthly visits provided an additional point of personal contact between Singer and its customers. Hata Toshiyuki in 1903 attributed Singer's global success in large measure to the installment plan: "The installment sales method is as follows. In the case of anyone who wants to purchase a Singer machine in one of these locations [around the world], the company will accept a small partial payment at the time that it gives the machine to the customer, who will pay the remainder in small weekly or monthly payments over a period of several years, finally taking full legal possession of the machine when the payments have all been made. This is said to enable those with limited resources to buy machines and use them."[48] That Hata felt compelled to offer this belabored account is a good sign of how little known the practice was at the time. Likewise, his use of the unconventional term *fukin* for installment selling suggests that no fixed vocabulary had settled into use.

At the time of Hata's writing, seasonal credit and various systems of revolving mutual aid credit (called *kō*) had a long history in rural society.[49] In cities, merchants allowed familiar customers to carry open-book credit that was settled monthly or yearly. But modern consumer credit—where the borrower (or her husband) earned a weekly or monthly wage enabling formally contracted "installment" payments—was a recent innovation. In the 1880s and 1890s, indigenous sellers in the lacquer industry, who for decades had sold on credit to rural households, shifted from seasonal to installment credit. These businesses evolved into "installment department stores" selling additional household goods such as furniture, bedding, *tatami*, and clothing.[50] Like the so-called "borax stores" that spread in the United States from the 1880s, these sellers appealed with shoddy products to relatively impoverished customers.[51]

Around 1900, foreign corporations, including National Cash Register and Encyclopedia Britannica, began to offer installment credit as well. Of these, Singer was the best known and most creative lender; it promoted installment buying to a new clientele of middle-to-upper-class urban families, women in particular. Unlike the native installment sellers, Singer used detailed written contracts; the cost of goods was high, the period for repayment long, and the credit premium modest. From around 1900 through the 1950s, a typical Singer machine represented about two months' wages for an ordinary salaried male household head. A machine selling for ¥112 cash in 1924 cost a total of ¥140 on a two-year installment plan, a roughly 12 percent annual rate of interest; installment department stores charged roughly 20 percent for goods sold on credit.[52] In Japan as around the world, installment sales made possible Singer's domination of the home market and constituted a critical practice in the making of the modern consumer.

The company waited until 1907, seven years after its full-scale entry into the Japanese market, before offering such credit. The reason for this delay is not clear. Singer perhaps needed first to accumulate sufficient funds in its Japanese headquarters and build a sufficiently widespread network of stores. But the company probably felt the need to cultivate a sufficiently disciplined sales force as well. Canvassers had to be taught not to offer contracts to bad-risk buyers. Collectors had to be engaged who would reliably turn over the proceeds of their daily rounds.

Although Singer's control of the hearts and minds of its sellers in Japan was always tenuous, installment sales proved popular from the outset. The company's sales reports show that machines bought on credit amounted to nearly 60 percent of all Japanese sales from 1909 to 1913.[53] Delinquency was never a major problem, but fragmentary evidence in these annual reports reveals an increase over time in the reliability of Singer customers. From 1909 to 1913, anywhere from 4 to 12 percent of accounts in Japan were recorded as one to three months' delinquent; longer-term delinquency (more than three months) never exceeded 1 percent and was in most years less than 0.5 percent. The only subsequent year for which partly comparable data can be found is 1928. At this point installment contracts accounted for about 63 percent of all Japanese sales, and even short-term (one to three months) delinquent accounts were less than 1 percent of all open accounts. Longer-term delinquency was not reported, probably because it was too low to bother noting. Over time, Singer's customers had become more disciplined in their use of credit, or the company more careful in whom they sold to, or both. Indeed, Singer arguably became too cautious, as domestic competitors—discussed in chapter 4—improvised new forms of credit that allowed them to find a market among less affluent, somewhat riskier customers.

To its Japanese customers, Singer presented the value of installment credit much as it presented the virtues of owning a sewing machine: as both convenient and a way to take part in a happy march of progress. A leaflet from about 1912 noted, "The past quarter century has seen all sorts of impressive progress in Japan. . . . [In every country] the sewing machine has completely revolutionized the household economy. . . . Singer Sewing Machine Company's monthly installment method makes it possible for any sort of family to purchase this great economic good. The Singer Model 28 Hand Turned Machine can be purchased for the incredibly cheap price of 3 yen per month, for 16 months, in other words 10 sen a day. In this way you can make a useful lifelong investment."[54] The purchase of a sewing machine on credit is thus neatly tied to national pride and progress, as well as individual economy and investment.

That Singer would everywhere in its advertising promote installment credit in such glowing terms is hardly surprising. Neither is it surprising that others viewed the purchase of goods with money not yet earned or in hand as economically or morally hazardous. The historian Lendol Calder shows that even (or perhaps especially) in the United States, early providers of consumer credit in the form of installment loans linked to the weekly wages of factory laborers or office workers were viewed with fear and scorn. As installment selling spread extensively in the United States from the 1880s through the early twentieth century, it "acquired a reputation for being the following of the poor, the immigrant, and the allegedly math-impaired female."[55] But Calder makes clear as well that even in these early days, not all forms of consumer debt were condemned; he describes a complex "moral topography" in which "some debts were considered justifiable, others not," and in which the key divide was between borrowing classified as either "productive" (good) or "consumptive" (bad).[56] His study also reveals a shift over time toward more tolerance and support for almost all manner of consumer credit.

Tokugawa Japan is well known for its widespread celebration of frugality and thrift, saving and investment, found in literary works as well as in exhortatory tracts aimed at women, farmers, or merchants, and in stringent government policies that condemned and sought to restrain consumption. Given this background, one might expect Japan's modern experience to be particularly tilted toward fear of consumer credit and condemnation of purveyors such as Singer. Certainly one finds evidence of such fears. The first systematic investigation of the modern practice of installment credit in Japan, *The Installment Selling System,* was published in 1929 by the Tokyo Chamber of Commerce. It reported on the United States, Britain, and Japan, noting that in Japan, as elsewhere, the early days of installment plans saw abuse in the selling of poor-quality goods at high prices and high-profile scandals in the installment sale of fraudulent securities. Installment credit naturally won a poor reputation among the "ordinary masses." Many had come to see purchasing this way as shameful and to see selling this way as below the dignity of reputable operations, such as the major department stores, which indeed eschewed the practice.[57]

One also finds linguistic traces of a critical view of installment selling. A pun reportedly bandied about from the early 1900s spoke of usurious consumer lending as ice candy (both were pronounced *kōri gashi*). Another alimentary linguistic play, this one said to have taken root in the 1920s, dubbed installment purchases as *"ramune."* This more complex play on words turned on the fact that the most common term for "installment sales," the word *geppu,* was a homonym for "burp." *Ramune* was a popular carbonated drink (the word is a Japanized pronunciation of "lemonade," which came into use in the 1870s) that caused the imbiber to *geppu.*[58]

But having combed newspapers and magazines from popular to highbrow in search of a culture war over consumer credit as the practice of installment selling began to take root in the first decades of the century, I find such voices of condemnation to be relatively muted. They are certainly no shriller than those raised in the United States. Women's magazines of these years often framed their articles in didactic and moralistic terms, but they generally echoed Singer's presentation of installment credit as not only a convenience but also a progressive and prudent activity serving individuals, families, and the nation.[59] They framed buying on time as part of a disciplined, economical, investment-oriented life. In January 1920, for example, a female columnist in the magazine *Fujokai* (Women's world) exhorted her readers to "increase the efficiency of the housewife or servant by boldly switching over to Western clothes for children. To this end, you must furnish the home with a sewing machine. In recent years this can be done with a small monthly installment payment, not that difficult for an ordinary family."[60]

Singer and its supporters in women's monthlies were anticipating and contributing to a more general understanding of consumer credit as a form of discipline and economy—not an invitation to excess or dissipation—that gained currency among economists and businessmen on both sides of the Pacific in the 1920s. In the United States, the pioneer advocate was E. R. A. Seligman, a Columbia University economist commissioned by General Motors to examine the pros and cons of the expanding and much-criticized provision of consumer credit. Despite the interested stance of his patron, Seligman's work, published in 1927, has stood the test of time as a classic statement of the positive value of consumer credit. He argued that "the installment plan induces the consumer to look ahead with greater care and to plan his economic program with a higher degree of intelligence. Many persons who would otherwise give little or no thought to the planning of future receipts and expenditures are virtually compelled by installment buying to construct what is in effect a personal budget."[61]

Two years later, the book-length report by the Tokyo Chamber of Commerce first affirmed the importance of "rationalizing" industrial management, a topic of widespread interest at the time among economic bureaucrats and the business elite. It then made a parallel case for "rationalizing" household management:

> In order to support the development of our national people's economy, it is necessary to reform the consumer economy, increase the efficiency of consumption, eliminate waste, lower the expense of daily life, and thus rationalize daily life. . . . [To this end, installment buying] is not only extremely useful in order to lead a disciplined life, planning a monthly budget of expenses; it also raises standards of living by allowing purchase of goods otherwise too expensive, and it is beneficial to expand sales. . . . In sum, skillful operation of installment selling will contribute in no small measure to rationalization of both daily life and [business] management.[62]

It is striking that these Tokyo authors, who read Seligman carefully and probably took a cue from his work in writing this passage, were even more forthright than he in stressing the rationalizing, disciplining function of installment credit. They placed this insight front and center in the preface; Seligman buried it mid-chapter toward the back of his book.

In several ways, then, from the start of the twentieth century, Singer brought to Japan an American way of selling and buying that increasingly reflected and reinforced values of a global modernity for men and women alike. The installment credit used for the majority of *mishin* purchases was framed less as a moral hazard and more as a disciplining force on both sides of the capitalist bargain. Buyers had to plan monthly budgets and keep to them; canvassers had to restrain themselves from concluding sales with poor risks, not only to help the company, but to insure that they—and the collectors—got their full commission. The women teachers not only helped their customers become morally upright home managers, with access to a world of convenience and efficiency, but they cultivated in themselves—and promoted to their students—a spirit of female self-reliance. The salesmen led this educational project, using science and system to enlighten the ignorant, lift the "pitiful" burden of housework on women, and create desire for the "happy sound" of civilization.

SINGER'S JAPANESE BALANCE SHEET

At first glance, Singer's first decades in Japan present a story of success. Its network of stores and its sales grew steadily from 1900 into the 1920s, and these operations were profitable. But even as the Singer way of selling and what came to be understood as an American way of life imbued both sellers and customers with the discipline and the desires needed to sustain a market economy, the company could not overcome one major impediment. Unless Singer could convince adult women—its largest group of users—to adopt Western dress or persuade them that Japanese dress was machine sewable, the appeal of this good was significantly limited.

Singer's household machine sales throughout the Japanese empire ranged from fifty thousand to eighty thousand units annually by the 1920s, with considerable variation from year to year (figure 6). Unfortunately no sources survive to explain how Singer or other observers understood this variation, but most of it follows logically from macroeconomic trends. The spike during World War I and immediately thereafter tracks the extraordinary wartime boom in Japan. The sharp drop in 1920–21 reflects the postwar recession. The spike in 1924 to the highest level of any year probably includes thousands of customers in the Tokyo-Yokohama region who were replacing machines destroyed in the devastating earthquake of 1923 during the post-quake "reconstruction boom." The dismal results of 1930 through 1933 track the world depression as well as the dispute that disrupted operations from the summer of 1932 through January 1933, and the gains of the following years (though

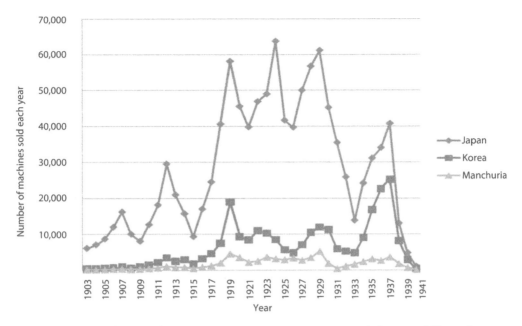

FIGURE 6. Singer sales in northeast Asia, 1903–41. (Japan War Loss Records, box 148, folder 5, Singer Sewing Machine Collection, Wisconsin State Historical Society)

we will see in chapter 4 that they could have been greater) reflect Japan's early recovery from the depression.

How profitable was this operation? According to Fred Carstensen, the company goal in the early twentieth century (at least for Russia, but apparently globally as well) was to keep the sum of in-country costs—store and office rents, commissions and wages, plus regional and central costs such as warehousing and advertising—below 45 percent of sales revenue.[63] The remainder was needed to cover manufacturing and shipping, as well as corporate headquarter costs and profits. Singer's annual world reports calculated total sales revenues and total expenses for each country (or colony) in which Singer operated. In many years, the preparers of this statement divided each country's or region's expense total into the revenue total, listing the percentage in a separate column; in years where this number is not listed, we can do the division ourselves.

It took some time for Singer to bring its Japanese operation up to its global standard. In its first two decades of full-scale operation in Japan, Singer almost always failed to keep expenses to 45 percent of sales. But

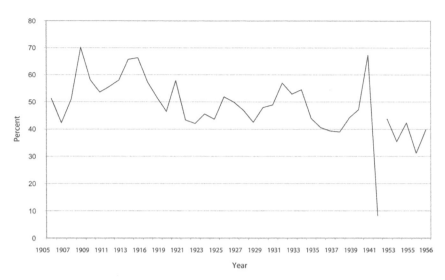

FIGURE 7. Ratio of Singer expenses to sales revenue in Japan. (Singer Annual World Results, 1905–1956, Singer Sewing Machine Collection, Wisconsin State Historical Society)

from 1921 until the eve of total war, with the exception of the worst years of local and world depression from the late 1920s through the early 1930s, Singer's Japanese operation generally produced results in line with these worldwide expectations (figure 7). Whether this mixed picture is viewed as successful depends on the point of reference. Carstensen notes that in Russia the 45 percent standard "was tough to meet." Singer began direct sales operations there in the 1880s, but "only after 1908 did Kompaniya Singer regularly approach or exceed" this threshold.[64] Similarly, Godley's comparative analysis of Singer's global sales results shows that the company rarely met this goal over the entire span from 1881 through 1914: "Over the period as a whole the costs of selling accounted for 58 percent of income."[65] By this comparative measure, the tried-and-tested Singer system worked as well in Japan as elsewhere.

But as Godley shows, the Singer system was built to maximize gross sales and market penetration more than profits, and by these measures the company could have done better in Japan. From 1903 through 1935, which was the first year in which a substantial number of machines were sold by Japanese makers, Singer sold roughly nine hundred and fifty thousand household-type machines in the Japanese home islands. Using an estimated 13.5 million households in Japan as the denominator, simple division yields a 7.5 percent household penetration rate. This calculation does not account for attrition—not knowable but certainly small given the extraordinary durability of these machines. Calculated not by household but as sewing machines per capita to align the data with Godley's analysis, which likewise does not account for attrition, cumulative sales, at roughly ten machines per thousand in the population (1 percent) by 1924 and fifteen machines per thousand by 1935, lagged significantly behind diffusion rates in the Philippines, South Africa, the Balkans, and the Ottoman Empire, and were roughly comparable to those for Ceylon. Singer's sales in the Ottoman region approached fourteen per thousand people by the start of World War I, two decades before reaching this level in Japan, despite the fact that per capita income in Japan in 1914 exceeded that of the Ottoman empire.[66]

In sum, although Singer's operations in prewar Japan were reasonably effective in controlling costs and returning profits, the company through the mid-1930s did not place an impressive number of machines into the hands of users compared to other medium-to-late developing markets outside the advanced core of the world economy. The explanation for this modest rate of diffusion begins with the fact that Japanese users—despite efforts to convince them otherwise—had come to see Western dress, Western sewing, and the sewing machine as a tightly connected—and either mutually reinforcing or mutually constraining—bundle.

These connections would be mutually reinforcing, of course, if Western dress carried the day. Imura Nobuko presented this logic in a women's magazine piece in 1920: installment payments made it easy to buy a sewing machine, in turn enabling the shift to more "efficient" Western dress for children.[67] And indeed, over the course of the 1920s, Western-style children's dress, especially among the urban middle or upper classes most likely to purchase a home sewing machine, became increasingly common. But until adult women made a similar shift in their understanding of efficiency and habits of dress for themselves, or changed their view of the use of the *mishin,* the market was limited. Singer made strenuous efforts to reject this bundling and argue that its popular family-type machine was useful for Japanese as well as Western dress. A 1910 ad in *Fujokai* dubbed the Singer Sewing Machine of "dual use for Western and Japanese dress." Eight years later in an "explanation of sewing machine use" accompanying the exhibit of a Singer machine at the Ministry of Education Exposition of Household Sciences, the company claimed that

> in sewing Japanese dress with a machine, as shown in the attached illustration, one can produce a stitch that can be taken apart with the same ease as a hand stitch, provided one practices using threads of appropriate type, with

the upper and lower threads different in strength and size, and one uses stitches of proper length and looseness, and needles of appropriate size. What requires particular attention to enable unstitching as in traditional hand sewing is to carefully distinguish the upper and lower thread, and practice to make sure that the lower thread is pulled with proper tension and practice drawing [the fabric] with the left hand. If one does so, the result is far superior to hand sewing.[68]

As this explanation recognized implicitly, *wafuku* had to be taken apart to be laundered, and its torturous complexity suggests the difficulty— indeed the impossibility—of convincing customers that they could easily produce the necessary loose stitching with a standard sewing machine.

In theory, Singer had an alternative path to reach Japanese customers: the chain-stitch sewing machine. The world's earliest sewing machines produced single-thread chain stitches, which would have been almost as easy to pull from the fabric as hand stitches. Although the lock-stitch machine won the day soon after its invention for most purposes and most markets around the world, chain-stitch machines are still produced both as a low-priced toy and at the high end for specialized purposes. Yet there is no evidence that Singer attempted or even considered marketing a chain-stitch machine for the home-based sewing of *wafuku*. One imagines that pride in the universal appeal of its family machine made the company unwilling to take this course. And Singer with good reason might have believed customers would recoil at the cost of owning two machines, one for Western and one for Japanese wear.

So Singer (and in later years its Japanese competitors) kept making the case for the dual use of the lock-stitch machine. A panel in an elaborate brochure from the early 1920s showed an elegant silk kimono embroidered with a Singer machine and displayed in the recent 1922 Tokyo Exhibition (figure 8).[69] But for the first three decades of Singer operations in Japan, few adult women either accepted the argument that the sewing machine could be used with *wafuku* or shifted to Western dress for themselves. Even those who supported Singer's effort to promote machine sewing of *wafuku*, such as the author of an article in the March 1919 *Fujin sekai* (Women's world) who sought to overturn the "misconception" that one cannot stitch Japanese dress with a sewing machine, could claim only that "ten percent of all Japanese dressmaking is done on sewing machines."[70] The sartorial and sewing preferences of women in Japan constituted a significant and continuing bottleneck to the spread of the *mishin*. We will see that this bottleneck only began to ease—and then only slightly—in the 1930s; only after the war did it give way entirely.

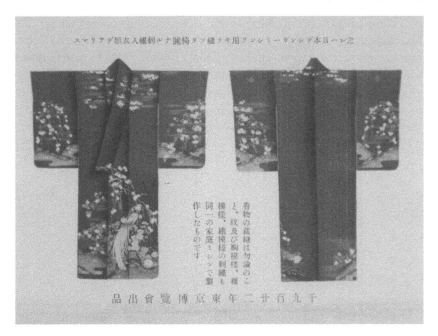

之レハ日本用ニシンミーガツシ用キラ縫タ輪絢ナル刺繍人女顎ノアリマス

着物の裁縫は勿論のこと、紋及び胸模様、裾模様、總模様の刺繍も同一の家庭ミシンで製作したものです

千九百廿二年東京博覽會出品

FIGURE 8. This kimono was put on display in a 1922 industrial exhibition and is
shown here in a panel of a Singer sales brochure. The caption reads, "Of course the
kimono itself, but also the embroidery on the collar, sleeves, and body can be done
with a sewing machine." (Edo-Tokyo Museum)

•••

As Singer set up hundreds of shops throughout Japan's main islands and
empire, the company made remarkably few adjustments to its selling system. Although the slow spread of Western dress among women limited
the machine's diffusion, Singer completely dominated the market that did
exist, so much so that women spoke of "my Singer" rather than "my
mishin." In that sense the company's one-size-fits-all global model was
a genuine success. Singer's experience in Japan—its rise through the early
1930s, and its subsequent fall—offers insight into the ways in which practices of global capitalism are not only transformative but also at times
resisted, and in some measure transformed, as they take root in particular locales. The dominant theme through the 1920s was one of trans-
forming impact. Singer played a role in shaping and promoting the modern profession of the salesman, new ideas of female self-reliance, and the
spread of consumer credit. But tensions with employees and their ex-

pectations, which surfaced in a number of small disputes through the 1920s, signaled a looming problem with the company's "universal" system of selling that would erupt in violent conflict 1932. In addition, the sewing machine would transform the fabric of daily life, especially for women, but it would take decades, through an uneven process of negotiated local meanings, for the *mishin* to become an object that virtually all families would both want and be able to possess.

CHAPTER 3

Selling and Consuming Modern Life

The Singer Corporation in the first decades of the new century established itself as a pioneer in selling mass-produced, brand-name goods in Japan. As it did so, its sales force and teachers and their customers, together with magazine editors, educators, and state officials who mediated their interaction, gave multiple meanings to the sewing machine. They linked it to changes in women's habits of dress and their roles in the family and wider economy. They discussed it in debates over women's contribution to progress and the modern nation. Various mediators were far from united in their views. Some sewing teachers argued that the disciplined practice of hand sewing would nurture feminine virtue in young women; other teachers and bureaucrats joined magazine editors to promote machine sewing as a means to individual self-sufficiency and national prosperity; still others, especially in the commercial press, saw the *mishin* as a tool for modern happiness. The dialogue among these figures reinforced a varied range of modern values and newly validated roles for women. Their negotiations also connected issues of gender with two further critical preoccupations of the era: anxiety over class division, and desire to make sense of a Western—most often an American—modernity while defining its Japanese incarnation.

MARKETING GOOD WIVES AND WISE MOTHERS

In print ads of the early 1900s, one core element in Singer's appeal remained the message first conveyed in 1902 by C. H. Pierce to Hata Toshiyuki: its machines would relieve the burden of household labor upon women "for the sake of the progress of the nation . . . [and] the advance of Japan's material civilization."[1] A Singer ad in the popular monthly *Fujokai* in February 1910 presented a young girl with a hand-cranked sewing machine, wearing kimono but with her hair tied by a Western-style ribbon. The accompanying text invoked this mantra of progress: "A twentieth-century family needs a twentieth-century sewing box. A twentieth-century sewing box is a Singer sewing machine" (figure 9).[2] This ad also trumpeted in large type a "Major Installment Sales" campaign, and a Singer leaflet from the same time likewise emphasized the centrality of a future-oriented economy in its appeal. The leaflet was ringed with ten-sen coins (100 sen = 1 yen), and the headline blared "Unprecedented Grand Sale" on an "installment plan of just 3 yen a month, 10 sen a day" (figure 10). The text stressed that "with the famous model 28 Singer sewing machine you can do all the family sewing for a LIFETIME."[3] The magazine *Fujin sekai* pounded home this economizing message in a Singer ad from April 1914: "Even *your* household finances can let you buy it. Our machine can handle all your family's sewing. And, it saves TIME, LABOR, MONEY."[4]

Two months later, a second ad in *Fujin sekai* put forth an important complementary theme in the marketing of the sewing machine: the appeal to pleasure. In this case the joy to be gained was that of keeping up with the latest in children's or women's fashion: "Now is the season to make summer clothes at home. The happiest and cheapest way is to use a Singer Machine."[5]

By the time of World War I, Singer had gained confidence enough in the persuasive power of its canvassers to forgo the cost of print advertising, but the brochures left at homes of potential customers allow us to glimpse themes in the company's direct sales pitches.[6] An early and interesting example from about 1912 presented a detailed and closely printed explanation of eleven reasons to purchase Singer's household sewing machine.[7] The initial appeal to "the study of home economy" brimmed with a familiar invocation of national power and pride: "In the past half-century all sorts of great progress has been made in Japan, and the progress of industry has been particularly outstanding. Japan is now one of the greatest industrial nations in the world." Singer expected the fruits of industrial advance to enter Japanese homes, for the invention and improvement in the sewing machine over "sixty-plus years" meant that "the household economy [had] been completely reformed by the sewing machine." In the West it was used by "all classes high and low." Global demand had risen to three million machines per year, and the best proof of Singer quality

FIGURE 9. Connecting the sewing education of young girls to a
new century in which sewing machines replace sewing boxes, Singer
with this ad appeals to the ideal of progress, even as it stresses that
its machine is suited to "dual use" for Japanese and Western dress.
(*Fujokai*, March 1910, page "i-no-ichi" in advertising frontmatter)

was the fact that Singer made and sold two-thirds of these. The claim that
Singer now operated more than three hundred stores and dozens of schools
and lecture courses throughout the empire told potential customers that
many of their compatriots had already heeded the message.

The third reason—"Appropriate for women of any status"—somewhat
cautiously added the cause of women's economic independence to the

FIGURE 10. With multiple images of a ten-sen coin, the daily cost of a typical installment contract, this Singer sales leaflet from 1912 stresses the investment value and affordability of the machines. (Nature and Science Museum, Tokyo University of Agriculture and Technology).

case for Singer: "For women who due to circumstance must make their way independently in society, the use of a Singer machine will bring greatly increased profit." The final point on behalf of the *mishin* restated the economic logic in more general terms. The installment plan meant that "Anyone can easily afford it." For example, "the amazing low price" of just ten sen a day buys the model 28 hand-cranked machine—"a lifetime useful investment." Although "amazingly low" was a stretch,

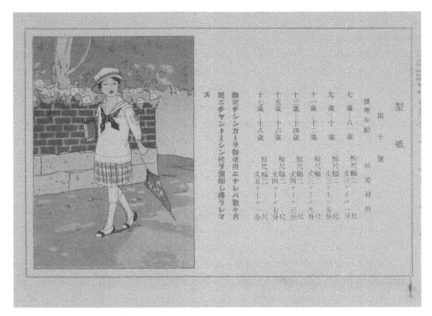

FIGURE 11. This panel from a Singer sales brochure, circa 1922, presents a stylish schoolgirl in what was likely her uniform strolling along a sidewalk in an elegant, brick-walled neighborhood, together with an appeal to the economy possible by "amortizing" the cost of the machine by fabricating these clothes at home. (Edo-Tokyo Museum).

such payments were not out of reach for the salaried middle-class. The monthly cost of three yen amounted to about two day's pay for a typical white-collar worker in his twenties. For at least some women of this era seeking economic self-reliance, this charge was manageable.

The use of a sophisticated economic vocabulary was one remarkable aspect of Singer's appeal to women in Japan. A brochure produced about a decade later included the following claim in bright red letters sandwiched between a picture of a young girl in a smart sailor's blouse and skirt, and a list of patterns for this outfit: "If you use a Singer in your home, you can tidily repay the cost in just a few months" (figure 11).[8] The key word here was repay (shōkyaku). Possible English renderings of this word range from repay or redeem to amortize. The ad writer chose a decidedly technical term from the several choices at his disposal to convey the sense of "make back" the money spent on the machine. The payback could come via two routes, left implicit in this brochure but frequently discussed in magazine articles: a woman might save money by making the family's clothes rather than buying them from a dressmaker or a department store (the more likely avenue in this ad); or she might earn money by producing clothes for neighbors or taking in piecework from brokers. Singer expected the modern women who were its customers both to accept the logic of rational investment for the future and to understand it as expressed in difficult language.

This brochure neatly mixed its appeal to rationality, discipline, and prudence with an invitation to a world of freedom and pleasure. The sailor blouse and scarf had been adopted as the uniform of many private schools for girls at the time, but the street on which the model strolled evoked an elegant European more than a Japanese cityscape. The adjacent panel offered patterns for young women's caps. As with the sailor suit, these were modeled by drawings of two Caucasian-appearing women, accompanied by the comment that "Western dress allows great freedom of movement."

The physical form of the Singer installment contract offered an additional invitation to participate in a prestigious practice of American origin. Although the vast majority of Singer customers, as well as almost all of the company's employees, and certainly all of the canvassers, were Japanese, Korean, Chinese, or Taiwanese, the installment contract was printed in English on one side and Japanese on the other. The two-sided printing was likely in the first instance a cost-effective convenience, making it unnecessary for a store to maintain separate contracts for Western customers living in Japan, and rendering the contracts legible to the expatriate top managers. But I suspect there is more to the matter. Although women's legal and economic rights in prewar Japan were highly circumscribed, surviving examples make it clear they could, and did, sign these contracts with their own names (with male guarantors co-signing).[9] As they inscribed their names in this contract and then used a machine with the English word *Singer* painted in bright gold on the arm, they were being told this was a made-in-America practice of economic modernity gone global, one in which they too could wisely and happily participate.

In such ways, Singer from the 1900s to the 1920s marketed the sewing machine in Japan, as elsewhere, as an emblem of modernity in two senses: that of rational investment on the one hand, and of freedom, style, and the pursuit of Western-linked pleasure on the other.[10] Given this globally consistent approach, it seems at first glance strange that the company placed an apparently traditional Japanese scene in the center of the same 1922 brochure that promised women they could "repay" the machine's cost while enjoying the "freedom" of Western dress. It depicted an extended

FIGURE 12. This panel, from the same 1922 sales brochure as the one shown in figure 11, seems to be selling the sewing machine by localizing it in a traditional extended family in a traditional room of tatami mats and shoji doors. Comparison to figure 13 reveals a more complex picture: this is a localized version of a modern family. (Edo-Tokyo Museum)

family, kimono-clad, in a tatami room with a formal alcove, gathered around a mother giving her child a lesson in sewing (figure 12). The illustrator seems to have been indigenizing this modern machine by placing it in a setting showing traditional dress, architecture, women's hairstyles, and a three-generation family, a scene in which a young woman was learning a time-honored feminine skill. But this is a misleading interpretation, for the source of this image was American (figure 13). It presents an iconographic translation of an early twentieth-century bourgeois American family ideal into Japan. This was a scene of "the first [sewing] lesson," used widely by Singer in its marketing from at least one decade earlier in the United States and similarly "translated" into other settings.[11]

The picture of the Japanese family has at its center an eminently modern figure, the "good wife and wise mother." As discussed in chapter 1, she was at this point fairly new in Japanese culture. These twin pictures literally illustrate the point that the good wife and wise mother in the early twentieth century was a modern global figure, not a traditional Japa-

FIGURE 13. This postcard, produced by Singer in 1913, depicts an American family of some means on the occasion of the (grand)daughter's first sewing lesson. In subsequent years Singer rendered this same tableau in other national or colonial settings of the non-West, Japan included. The company was exporting this image of the modern family together with its machine. Two versions of the card are found in the Smithsonian Archives Center's Warshaw Collection, Sewing Machines, box 3, folder 6. The dating is provided in Grace Rogers Cooper, *The Sewing Machine: Its Invention and Development* (Washington, DC: The Smithsonian Institution Press, 1976), p. 152. (Warshaw Collection, Archives Center, National Museum of American History, Smithsonian Institution).

nese one. The father sports a Western hairstyle and facial hair. It would have been most unusual for a father in a comparably high-status family of the Tokugawa era, living in this elegant home with art on the walls, to have been so closely involved in the supervision of a daughter's education. Likewise, the mother as educator was a new element of the Meiji ideal of "good mother." Finally, the high value placed on maintaining a close-knit "family circle" (*ikka danran*), nicely represented in this tableau, was a modern virtue that had come to prominence in Japanese discourse only since the 1880s.[12] This family circle was a means to promote both a modern life of education and discipline and a life where individual family members, including women, might find satisfaction or even pleasure through self-cultivation.

READING "TRUE STORIES"

What did women make of these exhortations to take part as good wives and wise mothers in the progressive project of making Japan modern? How widely would they come to desire or enact the American way of life put forward by Singer and its allies with its plural faces of prudent rationality, economic independence, and pursuit of pleasure? Would they accept it, stretch its boundaries, or otherwise transform or even resist it?

One rich source to explore for answers is the popular genre of the monthly magazine for women. In a managed forum, to be sure, these publications offer glimpses of women speaking for themselves. Usually edited by men, they featured articles by women and men, and frequent contributions from readers. The questions of how these publications were read, and how we might read them, merit discussion.

By the 1920s total monthly sales of women's magazines exceeded one million copies. The three most widely read magazines of the interwar era were *Shufu no tomo* (The housewife's companion), *Fujokai* (Women's world), and *Fujin kurabu* (Women's club). Each claimed a circulation in excess of two hundred thousand copies, and a single copy was often passed around among family or friends; the 1909 founding issue of a relatively short-lived magazine *Fujinkai* (also translatable as Women's world; figure 14) nicely conveys this with a cover showing three women, heads tilted toward each other, happily reading together a single copy of a magazine.[13] These publications certainly reached a significant proportion of Japan's adult women, especially in cities.[14]

These magazines built their popularity in part by interacting directly with their readers. *Fujin sekai* (yet another whose title translates to Women's world) ran a contest in 1908 in which readers were asked to mail in their plans for the best way to sew a hidden inside pocket into a kimono. More than three thousand readers were reported to have submitted designs. To judge the entries, the magazine engaged a panel of three experts (all women), including Hata Rimuko, the well-known principal of the Singer Sewing Machine Academy.[15] Such contests remained both popular and, for the winners, profitable. The July 1920 issue of *Fujokai* published four prize-winning essays on "How We Make Use of Waste Materials in Our Home." One "Sumiko" from Toyama Prefecture won first prize of seven yen for her account of using a sewing machine to turn her husband's old clothes into new outfits for their children.[16]

Such an essay fell into a genre labeled "True Stories" (*jitsuwa*), which figured prominently in women's magazines for decades. These first-person

FIGURE 14. This cover of the May 1909 inaugural issue of *Fujinkai* depicts three adult women, dressed in a variety of fashionable kimono, poring over a densely printed page of a magazine, inviting purchasers and readers to enjoy this publication among friends. (Courtesy of Tokyo University Meiji Shinbun Zasshi Bunko)

accounts by women whose experience was framed as exemplary, or cautionary, or sometimes both, offered one source of testimony from those who actually bought and used sewing machines. The magazine editors had the goal of selling and in some cases of enlightening. They chose stories they expected would attract readers, and stories that reflected their own views of proper women's behavior and values. Of course we cannot read the tales as transparent windows onto "real life." But the editors understood that telling a "true story" was a balancing act that required stories to be plausible accounts of a particular life situation, even if they were some distance from typical or average experiences. And running an essay competition required editors to feature topics that readers cared about.[17]

In addition to the strong response to contests, evidence that a large and ready audience of home sewers existed for true stories of needle-working women is offered by the monitoring endeavors of state officials and reformers. In February of 1921 the recently founded Social Bureau (*shakai kyoku*) of the city of Osaka examined the use of "free time" (*yoka seikatsu,* also translatable as "leisure time") in what appears to be Japan's first survey of daily life undertaken with modern methods of statistical sampling. Particularly (and, it would seem from the results, unnecessarily) concerned that youths and the working class might be poorly disciplined and profligate, the study questioned a random sample of nearly fifteen hundred Osaka residents drawn from categories including elementary school boys and girls (two hundred each), higher school girls (two hundred), male and female factory workers (two hundred each), as well as groups of "company and bank employees" and "merchants." The survey defined "free time" activity broadly to mean anything other than paid labor or sleep. It included housework and sewing. A simple questionnaire posed three queries: "What did you do with your free time yesterday? What do you usually do with your free time after work or on holidays? How much money do you spend on this activity?" These questions unfortunately do not allow direct comparison to later time-use surveys that counted the many minutes devoted to stitching. But these women did quite a bit of sewing. Higher school girls undertook sewing more frequently than any activity except "studying" or "reading." Among female factory laborers, it was the *most* common "free time" practice by a wide margin.[18]

Good evidence that magazine "true stories" struck women as believable and interesting is found in the "Survey of the Reading Habits of Working Women" carried out in 1934 by the reform-oriented Japan Library Association. Hoping to better grasp these habits so as to better guide women toward healthy magazine choices, the association surveyed employees at seven Tokyo department stores as well as workers in the postal service's insurance division. Of nearly sixty-six hundred women surveyed, close to three in four (72 percent) responded. Four of every five (81 percent) were young women, defined as aged 18–24.

The respondents were well educated: three in five had studied beyond the compulsory elementary level; two in five had graduated from Girls' Higher Schools. They were avid readers: the total number of magazines read exceeded the number of respondents. The three most popular were *Women's Club*, *The Housewife's Companion,* and *Fujin kōron* (Women's review), accounting for 73 percent of all the magazines read. *Fujokai,* reported to be in the top three in 1920, had fallen on hard times. Among the young women of Tokyo, it placed a poor seventh with only 102 readers.

All these titles featured similar articles. For our purposes, it is less important to measure relative popularity among them than to recognize that women clearly preferred gender-defined magazines over those aimed at a general (that is, a male) readership. Only tiny numbers read the most popular such publications, *Kaizō* (Renovation, thirty-four readers) or *Chūō kōron* (Central review, twenty-four readers). When asked which articles they preferred, women's responses were similarly gendered. The authors sourly noted that articles with "scientific" themes were unpopular, while "Music, movies, and entertainment" stories were most popular (17.8 percent). A close second was the category of "cooking, sewing and similar practical topics" (16.8 percent). Asked what sorts of articles they would like to see more of, 2,580 mentioned literature (both Western and Japanese), 2,252 mentioned sewing, and 2,193 hoped to read more on cooking.[19]

By following the sewing machine into Japanese homes and into women's lives through these magazines of wide circulation and appeal— and through parallel reading of sources such as the daily press and government or business reports—we learn much about the representation and the self-understanding of the good wife and wise mother. She was not only a new arrival on the social and cultural scene; she was also dynamic and complex.[20] Both in the discursive and social practices of daily life, women in early twentieth-century Japan took on multiple roles that were not always in harmony.

PURSUING PRUDENCE AND PLEASURE

For users as well as promoters, the sewing machine was intimately tied to values of economy and investment. A *Fujokai* reader identified by the given name Kikuko exemplified these virtues of the wise modern mother. Her thirteen-year-old daughter had learned to sew at school and wanted to continue at home. Kikuko, having earlier refused to buy a machine for herself, finally gave in to the inducements of a Singer canvasser, apparently paying cash in full, as she lists only the cash price (¥165) in her narrative. This woman of some means was nonetheless careful with her money. She expected the machine to pay for itself within a year and reported that with the help of the sewing teacher (presumably the Singer visiting teacher), her daughter started making her own clothes, indeed saving seven yen per outfit (about 70 percent by her reckoning) compared to a department store dress. The machine, she concluded, was both a good investment and a fine hobby.[21]

As articulated by serious-minded homemakers, consumer credit appeared frequently in a discourse of economy through rational modern sewing. Some women in fact anticipated a more general understanding of consumer credit as a form of discipline—not an invitation to dissipation or excess—only later put forth by economists and business advocates such as E. R. A. Seligman in the United States and the Tokyo Chamber of Commerce. Imura Nobuko wrote in her "true tale" of January 1920 that people felt the need to "revolutionize" family life. "First, one must increase the efficiency of the housewife or servant by boldly switching over to Western clothes for children. . . . To this end, one must furnish the home with a sewing machine. As this is in recent years possible for a small monthly installment payment, it is not that difficult for an ordinary family."[22] "Sumiko from Toyama," who won the essay contest on household economy with her piece on the reuse of old clothing, noted that "a *mishin* is very expensive, but there is a way to buy it with modest monthly installment payments."[23] A few months later, a writer from the town of Utsunomiya, not far from Tokyo, explained that one could now buy sewing machines on installment at three yen a month, save money on summer and winter clothing, find *tabi* patterns to make footwear, and also use the *mishin* to make lovely small accessories for one's children or for gifts.[24]

A "true story" published in *Fujokai* in 1930, one year after the Tokyo Chamber's book analyzing and advocating consumer credit, nicely affirmed its perspective. Aono Midori of Yokohama was married to a banker earning ¥100 a month, plus a ¥400 annual bonus. Mrs. Aono wrote with pride of economizing strategies that allowed her to get by on this pay. She had just bought a sewing machine and planned to save considerably by dressing her children entirely in Western clothes of her own making. Aono made very cautious use of credit; she purchased the machine in March with a ¥100 down payment and planned to pay off the rest in June with her husband's bonus, after paying installments of ¥5 in the intervening two months.[25]

Alongside such investment-oriented discourse, these magazines echoed Singer's promotion of the sewing machine as a source of pleasure and pride at taking part in a world of cultivation and fashion. This appeal appeared in magazine ads for other imported machines, and it played prominently in the ads of sewing machine schools as well. The Swedish maker, Husqvarna, in 1925 told readers of *Fujokai* that its machine was "a flower that blooms in the garden of cultural life," a good that would "visit your home with happiness."[26] The American maker Home Machine, in a series of ads in the early 1930s, offered "a convenient installment plan" and promised that "home users [would be] blessed with convenience and happiness."[27] Popular general works on the promise and peril of the installment plans that enabled women to own these machines echo this appeal. A snappily written book of 1930, *Sugu yaki ni tatsu geppu hanbai hō* (Immediately useful installment selling methods), grandly proclaimed that installment purchase would not only bring economic security to families by leading them to budget their expenses, but would also "democratize mass access to commodities, spread human happiness more equally, and relieve troubling class struggles." The "property-less intellectual class" would be able to afford a hundred-yen phonograph on installment but not for cash, and the practice thus "elevates the level of daily life."[28]

Compared to the blandishments of such advertising, in which appeals to rationalizing modernity were balanced by invocations of happiness, convenience, and pleasure, most "true stories" in women's magazines of the 1920s and early 1930s stepped lightly on the pleasure principle and stressed earnest attitudes grounded in prudence, planning, and investment. I doubt that the "truth" of these stories was the whole truth in the minds of women users; the sobriety of the "true stories" reflected the cautious and conservative moral stance of publishers and editors.[29]

Discussion of modernity and change in women's lives took place on a treacherous and contested terrain, which helps explain such a prudential editorial inclination. Consider, for example, this snarling blast at Japanese women in the May 1926 issue of *Women's Club* by one Watanabe Shigeru, a well-known educator in Japanese-style sewing and the principal of the Tokyo Women's Sewing School (forerunner of today's Tokyo University of Home Economics). In a piece titled "Whether to Invest in a Sewing Machine or Buy Bonds," Watanabe claimed that neither of the purported benefits of the sewing machine in the home, higher quality clothing or lower costs of production, were in fact enabled by its use. He staged a competition of "woman versus machine" to prove his point, setting five women to the task of sewing a kimono with machine and by hand. The machine was slightly faster on average, three hours and forty minutes versus three hours and fifty-five minutes, but it used twice as much thread. The *mishin,* he claimed, sowed "seeds of stupidity in the home." In Watanabe's harsh view, Japanese women did not think clearly or use their heads in daily life. They failed to see that the sewing machine was a faddish good wrongly touted as economical.[30]

On the front lines as a Singer salesman, Endō Masajirō was similarly troubled by the way women used—or failed to use—their machines. Asked by Singer to conduct a survey of owners, he was dismayed to learn that the sewing machines, often bought by women interested in embroidery, languished unused once the buyer's initial enthusiasm for decoration waned. The owners had not learned to use their machines for other purposes. Endō later recalled that only 3 percent of those he surveyed were actively using their machines. If women wasted money on expensive machines put to no good use, he worried, "it will only enrich the American company, and the spread of sewing machines will do nothing for our country."[31]

The next step in Endō's story rings apocryphal, but it is important nonetheless, for it circulated widely at the time and later. Moved by this concern, Endō joined hands with Namiki Isaburō to found one of the most successful sewing schools in prewar (and postwar) Japan, the Bunka saihō jogakuin (Cultural Sewing Academy, later renamed Bunka fukusō gakuin). In his view, he thus aided the nation by educating women to actually use their investments. Watanabe likewise took a pragmatic approach to the problem of poor investments by women; after completing his diatribe, he conceded that in some cases, if used wisely and effectively in the home, a sewing machine could be a reasonable purchase. But too many women bought it faddishly to keep up with the neighbors, believing that without a sewing machine one could not enjoy the prestige of a so-called "cultural life." They wasted 160 yen on a stupid purchase that went to little use. They would have been better off using the money to buy a bond; that was the action of a really civilized woman. For him, this was not a problem of individuals, but "a major national problem."[32]

By "nationalizing" the question of female wisdom or stupidity, Watanabe and Endō reveal that discussion of women and sewing machines took place as part of a contest to define what it meant to be modern— certainly a prudent investor, and for some a seeker of happiness and a cultured life. But the discussion also fueled the debate over how to be properly female and properly Japanese, and how to sustain social order in a changing world. Miriam Silverberg, in her wide and deep reading of *The Housewife's Companion (Shufu no tomo),* usefully divided its articles into three types, each of which, she argued, "challenge the official ideology pronounced by the state and taught in the schools": pieces on the choice of an ideal mate; reports on discord in the home; and articles "sending the modern Japanese woman out into the world (while at the same time encouraging her to bring the modern into her home)."[33] The "true stories" of *mishin* buyers and users fall squarely into the third category: articles on the pros and cons of homework (*naishoku:* the home-based production of goods sold to brokers or to neighbors); on the merits of sewing education and women's education more generally; and on the merits of Western versus Japanese dress. As they brought modern life home and connected home-based sewing to a broader world, the women in these stories indeed stretched the orthodox understanding of the good wife and wise mother as comfortably serving others—family and nation—while denying self. But the mainstream ideology of the family proved capable of stretching without snapping.

SEWING FOR SELF-RELIANCE AND SOCIAL ORDER

The 1920s witnessed in Japan a flowering of cosmopolitan culture marked in most accounts by jazz, Hollywood movies, and the "modern girl" and typically linked with the political trends of "Taishō democracy": cabinets dominated by elected politicians and their parties from 1918; suffrage expanded to all adult males in 1925; demands by women for suffrage; movements of organized workers and farmers. This linkage makes sense. Alongside the politics of parliament and party, the character of daily life was a central part of this era's cultural ferment.

Modern life as celebrated in media of the time was typically seen to originate in America. It was sharply gendered, populated by "salaryman" husbands archetypically earning their living as employees in state bureaus or the offices of Mitsubishi or Mitsui zaibatsu. Drawing at least equal attention was an array of modern life roles for women. These included the privileged full-time housewives of the urban middle class who engaged in formal study of cooking and sewing and sought the latest fashions on sale for themselves or their children in the multistory department stores recently erected in the major cities. One also found a small but growing band of full-time working women with jobs as dressmakers or hairdressers, typists or teachers. But the greatest attention—and no little criticism—were devoted to the character known as the "modern girl."

She was photographed strolling on Tokyo streets in an eclectic mix of Japanese and Western dress, lampooned in cartoons, lamented or celebrated in social commentary, and depicted in literature. The heroine of Tanizaki Junichirō's popular novella, *Naomi*, was a café waitress, and in his story, as in the media telling more generally, waitresses flaunted their sexuality and broke convention by choosing their own partners. In that regard, they were seen as more dangerous to society than prostitutes; by the end of the 1920s the waitresses outnumbered licensed prostitutes in Tokyo and had become objects of intense concern among the authorities and middle-class reformers.[34]

In this context of a widespread sense that change was everywhere and that change for women was particularly volatile, one finds discussion of sewing and sewing machines, among many other goods and activities, bound up in discourse over the perils and promise of modern life for women, their families, and the nation. This debate sometimes put female self-reliance, social order, and social class front and center, and in other cases foregrounded competing views of what it meant to be Japanese or live a properly Japanese life.

The question of homework (*naishoku*) for women was one issue that put into clear relief the dilemma of securing a stable order of both class and gender. Homeworkers amply testified to their efforts in numerous magazine articles, with *The Housewife's Companion* (*Shufu no tomo*) particularly keen to tell upbeat stories appealing to the entrepreneurial dreams of its readers.[35] One of the earliest came in October 1918, under the headline "Sewing Machine Homework Earns 45 Yen per Month: The Ideal Choice for Women Before Marriage." Author Fujiko, from Tokyo, explained that "after graduating from girls' high school, just to be prepared as the only child, I thought it important to have some occupation or skill, so while I hadn't yet lost that student spirit, I attended classes in machine sewing for three months at the Ueno branch of the Singer Sewing Academy." Three months was too short to learn usable skills, so Fujiko returned for a second semester, and then started looking for homework to put her skills to use. She distinguished between poorly paid subcontracting work, put out through several layers of brokers, and work taken directly from a large wholesaler. With the help of her father, who "kindly searched far and wide," she started to work directly for "Mr. Fujitsune in the Hasegawa district of Nihonbashi Ward."[36]

Fujiko described in close detail—the names and districts lending an authentic feel to the story—how she switched to a different broker whose jobs could be done by machine and then learned to satisfy his expectations for quality. She worked at the high end of the homework hierarchy; the silk material was quite expensive (said to be worth ¥200 per batch—about two months pay for a mid-level manager in a large corporation), so the wholesaler had to trust the homeworker not to steal or abscond. Introductions were crucial, and one left a "rather considerable" security deposit with the broker for good measure. To prevent "inconvenient" outcomes (presumably the cutting out of his middle role), he jealously guarded the identity of his customers from Fujiko and vice versa. She emerges in her story as a savvy and skilled young woman with considerable fashion as well as business sense. After "no ordinary struggle" of one year (including the six months of schooling) to establish herself, working from 8:30 A.M. to 5:30 P.M. she was able to earn forty-five yen per month (comparable to the starting salary for a male office worker of her age).[37]

Six years later, a "true story" in *Woman's Club* offered a married woman's tale of success in a collection of essays presenting "Our household's method of saving." The narrator was the mother in a family of five. She was fortunate to have a husband, employed in "a certain government agency," who earned enough to cover their basic costs, so her earnings were all saved. Despite her pretentious pen name, "Mirror-of-Heaven Child-of-Light," the author reflected a modest, middle-class, maternal ideal. She was determined not to sacrifice either health or family harmony by working long into the night or ignoring her children. She usually limited her homework to three or four hours a day sandwiched between housework tasks, occasionally working five to six hours. In 1921 she bought a Singer sewing machine on a three-year installment plan, with five-yen monthly payments, and she joined what appears to have been a rather informal "Household Occupational Study Group" to learn to make children's clothes.[38] Taking in jobs from a broker to sew winter or summer outfits, she was able to earn ten yen a month without overworking, clearing five yen after the machine payment. She put this into postal savings, but on the advice of her husband, she had recently withdrawn the savings and interest of ¥200 and transferred the sum to a bond of the Korean Colonial Bank. These government-backed bonds earned 8 percent annual interest, which the couple viewed as a very good deal. Now that the sewing machine was paid for, she expected to reach her goal of saving ¥1,000 in five years.[39]

Japan's empire figures in the happy ending to this story of hard work leading to a prudent investment, and the community of readers indeed reached into the homes of colonial settlers. A *Fujokai* feature on female entrepreneurs from 1932 offered five "true tales." One of the two from the sewing business was about a young woman of modest origins living in Taiwan. Miyoshi Hisae's father sold umbrellas, and she ended formal schooling at the elementary level. She studied sewing after graduation, and in January 1929 went to work in a nearby Singer store, one assumes as a teacher. Her mother objected to her working outside the home, so she began to sew clothes for sale to neighbors. Her reputation spread, and keeping up with fashion trends by reading the same magazine that published her essay, she was now earning about thirty yen a month. Miyoshi proudly detailed her expenses and income from a recent very busy month. At this rate, she would be able to pay for her own wedding.[40]

We must discount some for the advocacy in this story, which included the message that a woman should buy the magazine in order to prosper. In other "true" tales where the empire and a looming war figure prominently, the picture is darker. In 1934 Nakai Yaeko writes that her bureaucrat husband had been reassigned some years ago from a rural posting to the provincial city of Okayama. After several years, she joined him, when their children had reached middle-school age. Her family, like others in their village suffering from falling prices for their produce, had lost its land. Living an unpropertied city life for the first time, she spent many frustrating days scanning the newspapers for leads on homework.

Nakai owned a Singer machine, and eventually, in December 1932, she found needlework through the local office of the Women's Patriotic Association. A clothing store in town was under contract with the army to produce thick winter mittens. Her piece rate was ten sen a pair, and she was told she could expect to do twenty pair a day. But at first, she could sew at most two or three daily. It took a full month to finish her first twenty pairs. Just as her speed increased, she injured her fingers and spent most of February commuting to the local hospital. Soon after that, the work dried up. But a half-year later, in September 1933, the store brought her a new supply. Reportedly because the work had to be done "very carefully, with the needs of the state in mind," the piece rate had quadrupled. A table detailed Nakai's total earnings and costs from September through February 1934, which yielded a net income of ¥135 for eighteen hundred mittens.

Nakai's vivid account described a demanding work routine: about ten hours per day sewing, three in the morning, three in the afternoon, and four at night. At busy times, toward the end of the year, she worked each night until 3 A.M., but at the end of December she felt an "inexpressible satisfaction" at earning ¥37.5 for the month, a good bit less—and rather more plausible—than the fortunate Fujiko who had reported ¥45 monthly income in 1918 sewing kimono collars. Nakai reverently placed her wages on the Shinto altar in her home. She ended with pious words focused not on the state, but on gratitude for the opportunity to "gain hopes for my children's future."[41]

I have not found a single "true tale" ending with the complete failure of the homework endeavor. This surely reflected an editorial agenda common to the several monthlies that sold magazines by inspiring hope in readers. No doubt, many homeworkers faced more difficult conditions than Nakai, earned less money, and abandoned the effort. One imagines a body of readers—the wives of policemen, schoolteachers, railway clerks, or bankers—with a range of fears and desires. They feared that a calamity such as a husband being laid-off or running off, or falling ill and dying, would plunge them into poverty. They hoped they might stand on their own as "self-reliant" women. They dreamed or believed that with luck and hard-work, they might climb a rung or two into the promising new life of the middle class.

In such hopes, they were encouraged by the state and corporate world, albeit with real anxiety mixed with the exhortations. From the 1890s through the 1910s, observers had described homework primarily as a desperate survival strategy for the poorest of the urban poor.[42] But official investigations of homeworking in the 1920s, part of a new effort to scientifically assess and resolve social problems, began to qualify this dismal picture by noting that "the poorest of the poor in fact include relatively few homeworkers."[43] These studies made clear that by the 1920s, and probably earlier as well, many homeworkers considered themselves part of a growing new middle class and sought to bolster their buying power in that status.

According to a survey undertaken by Tokyo's Social Bureau in 1925, machine sewing was the best-paid and most desirable homework. The surveyors counted 2,638 sewing machine homeworkers, supplied by sixty-four different brokers and earning on average eight yen per month, significantly more than earnings from other homework jobs.[44] This and other accounts were ambivalent over the mix of possibility and peril in homework. The Social Bureau lamented that women, who "should be the heart of family life," neglected housework and the education and cultivation of children when they undertook homework. The homeworker's family "has absolutely no space" to enjoy a rich cultural life together, no time for "the pleasures of the family circle" such as "theater or strolls in the suburbs."[45] But even as the Social Bureau authors listed these problems, they noted that city officials were considering the promotion of homework as an economic relief measure. If homework was to be encouraged, these consequences must be addressed.[46] The family life threatened by homework, then, was that of the emerging middle-class of the imperial capital. The family-managing women in this bureaucratic mind's eye were expected to devote full time to household duties that in modern times included the "cultivation" of children and the orchestration of weekend outings. The problem was that homework simultaneously promised the wherewithal to take part in this life, yet threatened to destroy the cohesion and health of those families who did so.

Employers were less conflicted. In 1915, the Eastern District of Tokyo's Division of Communications, responsible for the city's telephone and telegraph service, issued a small booklet, "Encouragement of Homework," to lower-level male employees and their wives. The author lamented that a "mere" 494 out of 2,800 employee families were engaged in 107 types of homework, with sewing the most common.[47] The Communications Division sought to shame the other wives to help take the pressure off its budget. The author was frank: pay raises for the male employees were out of the question. Others had to take responsibility and do homework, overcoming a prejudice that this was a shameful sign of poverty.[48] The "benefits of homework" were manifold. It helped family finances; it stabilized the home by providing savings to cope with unexpected disasters; it encouraged diligence, turning wasted leisure time to good use; it taught wives the value of money—they were less inclined to waste money on lavish clothes if they had seen how hard it was to earn; it taught children the value of thrift, when economy-conscious homeworking mothers became less free with allowances, which were generally too high and wasted on things like candy and movies; it eliminated temptation to use idle time in decadent ways; it promoted family peace, as men and women became diligent and home-centered. And the nation benefited from a growing economy of made-in-Japan goods.[49] This illogical mishmash of claims saw as admirably home-centered the very practice that others feared would destroy the home. But if budgets were so tight, one wonders where these mothers had been finding the money to waste on shopping and children's allowances in the first place.

The author also played the Western civilization card, even as barbaric war raged in Europe. In countries such as Germany, it was no shame at all for women to embroider or weave. Their work supported the war effort, reinforcing national power and prosperity. Even those with considerable status or income must prepare for possible hard times, and "whether one is man or woman, household head or family member, rich or poor, we all have obligations as members of the nation to work to build that national wealth."[50]

This pamphlet was among the first of a steady stream of books promoting homework with grand exhortations that women build a strong national economy and reinforce family life, combined with practical advice on what jobs to seek and what pitfalls to avoid. The first peak of advocacy came just after World War I.[51] Homework promoters at the Ministry of Education, the national railway, and the Tokyo police were said to be urging the practice upon the wives of their many thousands of employees.[52] Sewing was the line of work best suited to women, and homework was expected to teach them to live more disciplined lives, something its advocates believed Japanese women were notoriously lacking.[53] Official and business analysts and advocates offered—sometimes in the same document—both a dismal vision of homework as a degrading practice corrosive of social order and morals, and a rosy view of it as a solution to social problems, a step on the path to independence and a better life for women and their families. On balance, homework figured in these discussions as a necessary practice, if not necessarily the best practice, to insure national and family prosperity and bolster social order. Homeworking women across the divide of working and middle classes had the opportunity and duty to support self and family economically, even as they managed the domestic sphere. Whatever the true difficulties of the homeworker's effort to balance the demands of productive and reproductive labor, the force of the exhortations of the commercial press, the state, and employers was to elevate the status of homework out of the realm of a shameful mark of poverty into a respectable, indeed an honorable, pursuit of the middle-class woman.

SEWING AND WEARING "JAPANESE-NESS"

Alongside and intertwined with these discussions of economic independence, social class, and social order, an anxious and gender-divided debate unfolded over defining and defending "Japanese-ness." Such concern reaches back well before modern times. But until the advent of Western power and culture as sources of both opportunity and threat, it was China that served as Japan's "unforgettable other."[54] Beginning in the 1870s, self-declared "Japanese" practices instead came to be positioned in a modern world where the West—and later, America—was the significant "other." One sees this in the 1890s among the young thinkers contributing to the journal *Japan and the Japanese* (*Nihon oyobi ni-honjin*); in the writings of Okakura Tenshin on Japanese, Asian, and Western art in the early 1900s; and later in discussion of an ever-broadening range of topics from home and family to work, education, food, and dress. Against this backdrop, promoters of a machine identified with both a Western mode of sewing and Western dress felt the need to stress ways in which the sewing machine was in fact suited to, indeed, supportive of, practices understood to be traditionally Japanese. Touching on topics small and large, they helped define these practices.

Hata Rimuko, in addition to founding the Singer Sewing Academy, wrote a best-selling textbook used at the academy and sold to several hundred thousand readers in more than two hundred printings of three editions from 1908 through the 1930s. In a surprising choice for a book focused almost exclusively on Western dress, the preface to the 1933 third edition noted that the author would retain the Japanese cloth measures rooted in Edo-era practice, rather than use Anglo-American or metric measures, even though the latter had recently been adopted as the nation's official standard. Because voices opposed to metric measures "remain loud," and "this is a matter of great importance for the progress of the national culture . . . for the time being we have decided to follow Japan's traditional measures."[55] In addition to reinforcing the idea that such practices constituted part of an essentially Japanese way, sewing machine manuals sustained stereotypes of traditional physical traits of Japanese women. Another top-selling sewing guide of the 1920s, authored by Souda Satoru and in its sixty-sixth printing by 1930, claimed, "It is truly a source of pride that Japanese women have such nimble and precise fingertips." Souda's endorsers echoed this theme: for *Jiji shinpō,* the manual "promotes use of the sewing machine in order to help Japanese women who are skilled in use of fingertips"; the *Osaka Asahi* added that the book "applies the particular skill in handcraft of Japanese women to machines."[56]

As further imagined by other leading figures in sewing pedagogy, Japanese-ness for women had meaning beyond skilled and nimble fingers. In 1903, Western sewing had been introduced into the government-approved sewing curriculum of public and private higher girls' schools, although these classes primarily taught Japanese dress and sewing techniques. Just a single unit in the third year was devoted to "using a sewing machine," and one unit in the fourth year taught sewing of shirts and pants. For more than forty years, until after World War II, there was almost no increase in the time allocated to Western sewing in the home economics curriculum. But, as interest in Western sewing and dress increased out-side the schools, educators in books, journals, public lectures, and gov-ernment-organized projects of reforming daily life engaged in complex, at times quite fierce, debate on the proper goals of their pedagogy. One axis of opposition was of particular relevance to the defense and defini-tion of Japanese-ness. It set those who saw sewing education as a core element in the moral cultivation of women against those who saw its mis-sion as the practical teaching of marketable skills.

Some twentieth-century educators who stood with sewing as a moral mission understood it to have deep roots in Japanese-style sewing. Ushigome Chie, an important teacher whose career stretched from 1910 to 1960, recalled that "in the provinces, until that time [1910s–20s], the idea remained widespread that skill in sewing determined a person's value as a woman. I was bad at sewing and calligraphy as a child, and was scolded at home, told 'You're not a girl.' This was not simply a judgment on skill in sewing, but a view of education that believed morality was nurtured through mastery of the techniques of what one might call the Way of Sewing."[57] Despite this background, Ushigome was more reformer than traditionalist in her pedagogy. She studied home economics in the United States from 1928 to 1930. Learning that sewing education there was part of a broader curriculum in "clothing" education, which taught everything from design and fabrication of clothes to socially appropriate and hygienic habits of dress and economical practices of purchase and care for a wardrobe, in 1931 Ushigome began to advocate (without immediate success) expanding the mission and renaming the sewing track as the "clothing track." The key point is that even as a reformer, Ushigome un-derstood sewing as a skill taught for the fundamental goal of cultivating "an abundant love of family" in Japan's good wives and wise mothers.[58]

With this view, Ushigome stood on the home-focused side of a split setting household against vocational sewing that had institutional as well as pedagogical roots. The institutional divide was not absolute, but in the 1920s newer commercial sewing schools were founded that placed relatively more emphasis on teaching sewing, especially Western sewing, as a trade for women. Such schools differed from both public and pri-vate girl's schools (elementary and higher) and from older commercial sewing schools, all of which primarily taught Japanese style sewing and saw it as a "Way" to be mastered by women in their domestic role as wives and mothers.[59]

Narita Jun was born in 1887, one year after Ushigome, and from the 1920s through the 1960s, she played a major public role as sewing educator. She studied at both the Kyoto and Tokyo Normal School for women, then spent two years studying in England before returning to teach at Tokyo Normal School. In the early 1920s, when she realized her students were fascinated with Western fashions, Narita undertook to instill a love of sewing by teaching them the "pleasure" of designing and fabricating their own Western-style clothes. But even as she promoted Western sewing and Western dress, Narita believed that the value of sewing rested in its place as a home-based, womanly art.[60]

Advocates of this position, such as Narita and Ushigome, were not reactionary traditionalists defending Japanese-ness against the onslaught of the modern West. Narita in the 1920s both taught Western dressmaking and aligned herself with educators who continued to justify sewing education as a moral rather than a practical or occupational pursuit. These colleagues drew upon liberal Western theories of pedagogy as they stressed the need to cultivate the whole person in educating good wives and wise mothers. Among their leaders was Kinoshita Takeji, a generation senior to Ushigome and Narita (b. 1872). Author of books on pedagogy in general and sewing education in particular, he spoke from an authoritative perch as professor at Nara Normal School and Girls' Higher School and principal of the attached elementary school. He shared with Ushigome and Narita the desire to go beyond a focus on skills to teach "a spirit of sewing" (one might also translate his term, *saihō gokoro*, as "a sewing heart"). For Kinoshita, this spirit centered on values of freedom as well as cooperation. The "sewing heart" had an intellectual dimension, promoting powers of memory, observation, and imagination; an emotional dimension anchored in appreciation of beauty as well as confidence in oneself; and an "intentional" dimension of promoting resolve and motivation. Sewing for Kinoshita included Western sewing and the use of sewing machines, but it was never to be limited to occupational training.[61] He and his colleague advocated a moral education emphasizing individual-oriented ideals of citizenship.[62] Their pedagogy affirmed that a woman was to be a good wife and wise mother above all. But they saw this as a defining virtue not of a Japanese culture that resisted modernity, but of one that embraced it.

Outside of the classroom, efforts to define or defend Japanese identity in daily life and consumer choices were often directed at the threat or promise of what came to be called the "two-layered life" (*ni-jū seikatsu*, also translatable as "double life"). The term refers to the simultaneous presence of goods and practices described as "Western" and "Japanese" in realms of food, housing, and dress. Newspapers, magazines—especially those for women—and more academic publications for professionals in fields such as architecture and home economics were full of anxious discussion of the pros and cons of practices such as tatami-sitting versus use of chairs, and of course the merits of Japanese versus Western dress.[63]

The Japanese state acted to frame this discussion of the "two-layered" life and daily life more generally. In the immediate aftermath of World War I, as the state broadened its effort to reinforce order in a contentious society, bureaucrats allied themselves with middle-class reformers and experts in various fields, sewing and home economics among them. Officials in the Home Ministry and the Ministry of Education played the key role in founding one such initiative in 1919, the League to Reform Everyday Life. Over the following years, in publications, exhibitions, and lectures, the league connected state officials with outside experts and activists in committees concerned with rational and scientific housing, food, and clothing.[64]

The key reformist group in the realm of sewing was the Japan Dress Reform Association. Founded in 1921, the Association immediately published a book written by two of its leading members, the husband and wife team of Ozaki Yoshitarō and Gen. Titled *Sewing for the Future: Economic Improvement,* and aimed at women readers, this massive tome of nearly five hundred pages sold seventy thousand copies.[65] Its aim was to lift the burden of the "two-layered" life by (literally) reforming Western dress to better fit Japanese bodies. For such a dress to spread, it had to be simple enough to fabricate in the home, with or without a sewing machine, and it had to be economic in its use of fabric. It also had to allow free and active movement. Among the ideas for such functional, hybrid dress design taken up by these 1920s daily life reformers were modified versions of rural Japanese women's work clothes, including pants known as *monpe.* Stillborn at the time, these initiatives would be revisited in the late 1930s, in a similar rationalizing spirit.[66]

Hata Rimuko echoed this concern with the two-layered life in the 1924 preface to the second edition of her sewing textbook: "Recently, the numbers of women and children wearing Western clothes has increased greatly. This appears at first to be a welcome trend, but if this is simply a matter of pursuing fashion with no sense of principle, it is most regrettable. The burden placed on the nation by the spread to women of the two-layered life in clothing is fearful even to think about."[67] She continued by noting that this burden could be removed "if it were possible simply to cease making Japanese dress," but she admits that "far into the future, women are not going to abandon Japanese dress." Given this situation, she offered some options to ease the economic pain: limit Japanese wear to ceremonial occasions; only gradually move from Western dress for work to Western dress for social events; or adopt those Western fashions such as the "one piece" dress that are similar in fit to Japanese clothing and easy to sew.

Hata had for years been fighting a sort of culture war against what in 1907 she had called "the stupid idea that machine sewing does not harmonize with the Japanese home."[68] She was well aware that Western dress was more amenable to sewing by machine than Japanese dress, or at least that users believed this to be true despite Singer's marketing of the *mishin* as "dual-use." Her interest as head of the Singer Sewing Academy and as wife of a highly ranked Singer manager was unambiguous: the more quickly women resolved the double-life issue by moving to Western dress, the better. Her cautious and defensive preface despite this interest speaks eloquently to the way dress for women in particular carried a burden of defining cultural identity in a changing world. In 1924 she was still struggling to harmonize a world of dress options understood in terms of nations and cultures, a world in which a dramatic increase in the use of sewing machines by dressmakers or housewives awaited the further adoption of Western dress by Japanese women, while adoption of Western dress required willingness to embrace machine sewing.

As Hata grappled with this dilemma through the 1920s, some of Japan's best-known writers, such as Tanizaki Junichirō, were writing with nostalgia about the disappearance of what they saw as traditional lifeways because of the uncritical embrace of the modern and the Western.[69] Both literary works and Hata's commercial advocacy rested on a nested array of commonsense binary oppositions: Japanese sewing, hand stitching, kimono, and "tradition" set against Western sewing, machine stitching, Western dress, and "modernity." But as in the case of debate over sewing education or efforts to reform daily life more generally, it is profoundly misleading to view the debate over sewing and dress modes as a simple fight between "Japanese" tradition and "Western" modernity. Almost all parties were making claims for the practices best suited to a life understood to be both Japanese *and* modern, one that embraced values of speed and efficiency, freedom of movement, and sometimes even freedom of choice.

Consider, for example, the impassioned 1921 presentation by a teacher at a Girls' Higher School in Kobe of her time-saving technique for housewives: hand stitching that was faster and more efficient than machine sewing. Echoing the Osaka study of free-time use that same year, she noted that Japanese women spent inordinate amounts of time on housework, above all on sewing. This made them backward compared to Westerners, and the long hours spent to learn sewing in school impeded study of other subjects. Her solution was a unique stitching technique for "hand sewing faster than machine sewing," which she sought to validate with contests in which her students sewed coats and kimono sashes by hand more quickly than veteran sewing machine operators.[70]

Among schoolchildren as in adult society, the move to Western dress took place on a gender-divided time line, but because school girls began to wear Western-style "sailor" uniforms in large numbers decades before adult women donned Western blouses and skirts, their choices provoked the earliest iterations of a long-running debate on nationally defined cultures of dress. Beginning in 1886 the Westernizing Minister of Education, Mori Arinori, required middle-school boys to practice military calisthenics. With this new bodily movement came a move to a Westernized, military-style schoolboy uniform worn with virtually no alteration in public schools through World War II, and still worn in some private schools today. Schoolgirl dress followed a more meandering and controversial course. The spread of calisthenics from boys' classes to the future mothers of the nation, especially in the decade from the first Sino-Japanese War (1895) to the Russo-Japanese conflict (1904–5) generated pressure to reform dress. But the initial push from the government was not for Western wear but for "girl's *hakama*," a modified version of the traditional male skirt of Tokugawa times.[71]

Although the *hakama* spread widely to become an emblem of the modish schoolgirl at the turn of the twentieth century, the push for this mode produced heated discussion. Reformers spoke of the need for dress to allow ease of movement in a new century that would call on women to compete as men did. Opponents feared an outbreak of decadence or assertiveness.[72] Reformers won the day, apparently because the schoolgirls themselves were delighted with this new wear. For supporters, the *hakama* both liberated women's bodies and was aesthetically pleasing. The skirt carried a class marking together with this newly gendered one, that of the privileged status of "girl student."[73]

As this uniform school fashion began to take root in the early 1900s, the early monthly magazines for women, such as *The Lady's Companion* (*Fujin no tomo*) and *The Household Companion* (*Katei no tomo*), began to promote Western dress for children to their middle-class readers with photographs and instructional columns.[74] Over the next two decades, lively discussion took place among teachers, students, school administrators, and state officials over the merits of continued modernizing reform of Japanese-style dress versus the shift to Western dress. Discussants agreed on basic goals: the modern virtues of convenience, hygiene, economy, and mobility, as well as the importance of cultivating women's aesthetic sense. During World War I, the Ministry of Education kept its sights on reformed *hakama* with short hems to allow easier movement, and similar calls for modern but Japanese-style schoolgirl dress persisted under the umbrella of the postwar campaigns to reform everyday life.[75]

But eventually, by the 1920s, nearly forty years after boys put on uniforms based on European army dress, the schoolgirl "sailor" uniform won the day.[76] This new style reached a sharply increased body of students, as the number of young women going on to post-elementary Girls' Higher Schools leaped from about 90,000 in 1914 to 326,000 by 1926. Not only teachers and parents, but students themselves took part in these decisions, which were not imposed by fiat, despite the participation of education officials. Some students pushed schools to implement Western-style uniforms and were denied; others argued that uniforms destroyed women's ability to make aesthetic judgments. Young girls at the Eastern Girl's Academy (Tōyō Joshi Gakuen) pressed teachers to implement Western dress uniforms and were scolded for their worship of Western fashion and betrayal of their "Eastern" roots. This shift to the sailor uniform was a diverse, contentious, but surprisingly open process involving study groups and committees that included parents, teachers, students, and school administrators.[77]

As the dust was settling on this issue in the 1920s, sustained debate over the merits of Japanese versus Western dress for adult women was just beginning. Despite Hata's anxiety in her 1924 preface, the adult women who were the principal users of the sewing machine, far from embracing a double life, were mostly wearing kimono. The famous ethnographer of daily life in Japanese cities, Kon Wajirō, undertook the first of his surveys of dress in public places in 1925 by counting women on the streets of the Ginza district of Tokyo.[78] He found that just 1 percent of women were wearing Western dress; clearly the anxiety in the dress debate of these years was a reaction to the prospect of change more than its achievement. What is fascinating is that the values of "modern" life—efficiency, economy, and convenience on the one hand, and freedom and pleasure in a disciplined framework on the other—underlay both sides of the argument.

In an early example, an April 1923 forum in *Women's Club* addressed the "pros and cons of Western and Japanese dress." Advocates among the eighteen prominent men and women consulted saw Western dress as best for ease and speed of activity, especially for students and children. They described it as especially suited to public activities and places, and as more economical in both time and money. The minority of dissenters argued from the same economic logic: the financial burden of supporting the "double life" ruled out incorporating Western dress into the wardrobes of ordinary people. One participant also mentioned the "poor physique" of Japanese women as a reason to change slowly or not at all; but here too, the long-term goal was to "improve" Japanese female bodies in the direction of Western physiques, so women could someday make this change.[79] A few months later, the tragedy of the Kantō earthquake of 1923 prompted related claims for a safety advantage to Western dress: women in kimono were dangerously slow in an emergency.[80]

Fujokai joined the fray the following spring with a lively panel on dress for working women. It came down on the side of Western dress, and for similar reasons: ease of movement, ease of laundering, savings in time and money. One educator brought a note of humor to the table, noting that her school faced increased costs for broken windows after the switch: freer movement in gym class led to harder throws and more balls crashing through glass windows! She also won agreement from the others in claiming that the long hours devoted to the teaching of Japanese-style hand sewing were deplorable; girls lagged behind boys in their studies because they spent six to eight hours each week in sewing class. By teaching machine sewing (for Western dress), she could cut class time by two hours per week, freeing time for more crucial subjects (unfortunately not specified).[81]

Voices on the other side affirmed a modern logic, noting that whatever the merits of Western or Japanese dress in theory, efficient use of time or money depended on a woman's practical familiarity with one or the other mode. Those brought up on Japanese dress found it easier to make and to wear. One advocate of Japanese dress also, obliquely to be sure, promoted it by invoking the value of pleasure. He noted that men found women in kimono more glamorous or attractive; while others might think that Western dress was more becoming, he argued that women in any case—so long as they did not sacrifice their "character"—should be enabled to take pleasure in their appearance, to enjoy looking and dressing beautifully while young.[82]

Anxiety about the way dress might subvert social order by promoting sexuality in the wrong contexts was clearly on the minds of the men

at the roundtable. One opined that Japanese kimono were simply too seductive: they drove men crazy; they were not safe for women at work.
But another countered that skirts could be just as seductive.[83] Of interest in this exchange is that the concern of both men was to encourage
dress that would bolster order and, one assumes, productivity in the workplace, even as it might offer a certain pleasure to both men as observers
and women as stylish dressers.[84] Perhaps the strangest case where concern with women's bodies and dress choice, the sexual gaze of men, and
the value of efficient movement intersected in public discourse was that
of the Shirokiya department store fire of 1932. In the wake of this dis-
aster, which claimed thirteen lives, including seven women, the perhaps
apocryphal story circulated that their Japanese dress caused the female
employees trapped on the upper stories to fall to their death. The story
went that because a kimono was typically worn without underpants, the
women feared exposing themselves to the crowds gathered below. With
one hand holding their kimono hem in place, they had only one hand
to grasp the escape ropes hung from windows and therefore lost their
grip. After the tragedy, fire officials and a Shirokiya executive reportedly
advised female employees and visitors to high-rise buildings to wear
Western-style "drawers," and some stores were reported to be requiring
this of employees.[85]

Despite these widespread, at times sensational concerns that dress
should facilitate productivity and safety, actual change in what women
wore to work, much like change in their school uniforms, took place at
a much slower pace than for men. By the late nineteenth century, most
male factory and office workers wore Western pants and shirts. But even
in the most modern of the textile mills, which employed hundreds of thousands, women's wear only slowly shifted to Western styles. While employers or reformers were surely interested in efficiency, they struggled
mightily to balance this objective with their understanding of women's
cultural roles and obligations. An early example is found in 1917 in Japan's
first labor union publication aimed at women workers. The author wrote
of the need for hygienic and safe work dress. It must keep the body warm
and clean and protect workers from injury, but it also must allow women
to present themselves appropriately in various contexts, from work to
more formal occasions. All her examples were variations on Japanese-
style (*wafuku*) dress.[86] Similar discourse continued into the 1920s. A pioneer in the field of labor science, Ishihara Osamu, in his 1922 book *Rōdō
eisei* (Labor hygiene), noted that Western dress was preferable to Japanese wear, but, aware of prevailing trends, he simply urged women in

any case to wear short-sleeved kimono to avoid snagging their arms in the machinery.[87]

Photographs from the 1920s into the 1930s in commercial sources such as *Asahi gurafu* (Asahi graphic) and union publications such as *Rōdō* (Labor) make it clear that through the end of the 1920s, most laboring women in ceramic work, tea sorting and packing, breweries, match factories, rubber factories, cosmetic factories, and confectionaries wore Japanese dress.[88] But gradually from the 1910s into the early 1930s, major employers did conclude that Western dress was the more productive, safer choice and that it could be made sufficiently pleasing aesthetically. Tōyōbō Spinning moved in the 1910s from tight-sleeved kimono to skirts with jackets, and in the late 1920s adopted a blue summer uniform likened to "flowers blooming in the factory."[89] The authors of a pamphlet reporting on a sort of industrial fashion show of 1934 claimed that interest in women's work dress had intensified since a meeting the previous year in Osaka on factory safety, efficiency, and dress. Although poor dress design was acknowledged to contribute to a significant number of accidents, the issue was said to be more complex than at first glance: for women, one had to design clothing that was not only safe and efficient, but aesthetically pleasing to preserve "feminine grace." The pamphlet included photographs and descriptions of women's work dress at eight factories, mostly in the textile industry. All now provided Western-style uniforms, with several noted to have made the change only in the past one or two years.[90]

As the dress debate broadened from schoolgirls and working women to all women, it moved from women's magazines and union journals to the national press. In 1925 *Tokyo Asahi* ran a six-part series, "For and Against Western Dress," with numerous contributions from readers and a roughly fifty-fifty balance pro and con. Affirmative voices, in a now familiar refrain, stressed economy and ease of movement—for Hirata Utako, of Ogikubo district of Tokyo, especially in getting on and off busses—while critics such as Uchida Akiko of Ushigome in Tokyo argued for Japanese dress as more economical in avoiding the burden of the "two-layered life": one did not need a sewing machine or any special training in Western sewing.[91] In addition to economy, the modern rhetoric of improvement was embedded in the arguments of kimono advocates. Kanazawa Tsuyako admitted that Japanese dress was to some extent inconvenient for the hustle-bustle of modern life, but she argued that this drawback should be addressed by improving, not abandoning, Japanese dress, which even foreign women were coming to appreciate.[92]

The one motif in this forum and elsewhere that directly opposed the logic of accepting modern values echoed Tanizaki's nostalgia by defending Japanese dress on grounds of the fit between the physique of Japanese women and the aesthetics of Japanese dress. But such voices rarely rested their case on this point alone. Takahashi Setsuyo, in *Tokyo Asahi*'s forum, asserted both that Japanese dress was best suited to Japanese women's bodies and aesthetic sense, and that it was quite well-suited to physical labor. The actress Sakuragi Fumiko likewise combined an economic with an aesthetic case for Japanese dress: Western dress was too expensive because it demanded too many accessories, and it was ill-suited to Japanese homes or the aesthetics of Japanese women.[93] The officials, the civilian experts, and the ordinary letter-writers who so vigorously argued over sewing pedagogy, hand versus machine stitching, or the merits and burdens of the double life in dress, accepted without question the values of efficiency and rationality, and sometimes those of freedom and individuality. They simultaneously affirmed the inevitable and irreversible advent of modernity *and* the impossibility of discussing it without addressing its Japanese-ness or its moral dimension.

. . .

Amid lively debate but a still limited move toward Western dress for women, the five hundred thousand household machines that Singer sold in Japan between 1903 and 1930 established a modest foothold in actual ownership and a much larger presence in the cultural imagination. Even with the help of low monthly payments, by 1930 no more than 3 to 4 percent of Japan's households owned these machines. Advertising, magazine articles, word of mouth, and the ubiquitous presence of the Singer salesmen made this good far more widely known and desired than owned or used. In this sense, the story to this point has been one primarily of cultural history—the spread of images that would frame the dramatic social transformations to follow.

Singer and its friends in the publishing world presented the *mishin* as an indispensable tool of the good wife and wise mother, and her daughter. Its purchase and use promised the opportunity to partake of modern life in its twin guises of future-oriented rationality and present-minded pursuit of pleasure. The *mishin* also offered women the prospect of achieving independent economic power. She was to be a rational manager of the family economy; a happy or even a profligate shopper; and a proud seeker of self-reliance. The sewing machine thus validated new roles for women in both domestic and public realms. Although these roles sparked debate and expanded the possibilities open to women, by the early 1930s they had not fundamentally challenged the power of men or the subordinate position of women in most households. This enlarged space rather helped forestall a more profound transformation of gender relations. This space of possibility also promoted the view of class as rungs on a ladder to climb rather than categories doomed to struggle.

The sewing machine also figured in vigorous debate over the place of Japan and Japanese-ness in the modern world. It is tempting to understand this debate as a set of battles pitting new technology and new practices identified with the West and modernity, such as the sewing machine, Western-style sewing, and Western dress, against older practices identified with tradition and Japanese-ness, such as hand sewing and Japanese dress. Although the sewing machine was sometimes positioned in this way, defenders of Japanese tradition most often and most effectively made their case by anchoring it in modern values.

Some looked to hybrid forms, taking the best from Japanese and Western dress styles to overcome the pressures of negotiating a two-layered life in clothing, but their advocacy had little impact on daily life. Overall, the dress debates only sharpened the sense of separation between styles marked as "Western" and "Japanese." To the extent that some women were beginning to move away from Japanese modes, they were not putting on the hybrid reform wear of expert design. Rather, the most popular dress trend for women in the 1920s was a simple, short-sleeved, tube-like summer dress, the *appappa*. The word's origins remain obscure; it perhaps derives from "upper part" in reference to the sleeveless top. The *appappa* spread first in the Osaka region, both by word of mouth and vigorous promotion by magazines such as *The Lady's Companion*. By the end of the decade it was a common sight, found as much in the lower-to-middle class districts of major cities as in fashionable suburbs.[94] Its appeal was noteworthy for its variety. It was a Western design, but unlike most dresses, it was simple to sew along straight lines, making it easy to sew by hand for those who did not yet own a *mishin*. It was simple to wash. It used inexpensive cotton fabric. And it allowed easy movement. Equally notable is that its path to adoption ran from the bottom up and from the outside in, that is, from Kansai toward Tokyo. The practices of modern life were both widespread and generated along multiple paths.

CHAPTER 4

Resisting Yankee Capitalism

From late summer of 1932 through the following winter, employees of the Singer Sewing Machine Company organized two labor disputes. The violence and the anti-American focus of the second made it news "fit to print" prominently in the *New York Times,* unprecedented for a labor struggle in prewar Japan.[1] Salaried employees led the first action. Some of them took part in the much larger second dispute, as did managers and sellers in some of Singer's provincial shops. But the main body of protesters in the second dispute was the selling corps of branch managers, salesmen, and installment collectors in Tokyo, Yokohama, Osaka, and Kobe. These men worked on commissions that were shrinking in a depressed business climate, but their grievances reached well beyond discontent at declining pay.

The disputes were notable for the leading role played by managers and sellers, for their tactics—a virtual takeover of the sales apparatus—and for the ethnic nationalist rage that fueled the fires of grievance. The employees challenged Singer's property rights, and they voiced widely held beliefs about the terms of employment that were legitimate in a capitalist system that was increasingly understood and desired to be specifically Japanese. Their action ended in total defeat, but in the longer run the company suffered the greater loss. In stubbornly defending the global validity of a single selling system, Singer paved the way for Japanese competitors who cloned Singer's product but more effectively adapted and legitimized its sales system for sellers and buyers alike.

PRELUDE

During the 1932 strike, Singer laid claim to a thirty-two-year history of harmonious relations with its employees and portrayed the protesting workers as a small faction of radical malcontents. The Kansai Employee Federation countered that "over the past thirty years of operations, nine disputes have broken out in Osaka, and over 20 have taken place if we include Tokyo and provincial shops."[2] In prewar Japan as in other times and places, the definition of a dispute was open for debate, and not all contentious moments were covered in the press or captured in official statistics. But the employee claim for a disputatious past is in part confirmed by documentary traces.

In March 1925 the Alliance for the Reform of the Employee System, based in Tokyo but drawing support from Osaka as well, demanded "lower prices" (of the sewing machines, one assumes) and changes in the employment contract. For the duration of their action, some 50–60 employees continued business operations but withheld revenues from machine sales and installment payments.[3] Lacking details of the predispute status quo, we cannot be sure what the employees gained in the fourteen-point "agreement" signed on March 28 by a Singer expatriate executive and six representatives of the employee organization. But it is clear that conditions of employment were seen as legitimate issues for negotiation and adjustment by both sides. The company set forth stipulations (whether new or unchanged is unclear) concerning the calculation of bonuses for installment collectors and similar matters, and promised "to study and consider the matter of retirement or severance pay" and "consider the wishes of the alliance concerning treatment of salaried workers."[4]

Nine months later, a group simply named Alliance to Reform the System launched a dispute in Osaka protesting Singer's decision to close five of the city's twenty-five sales offices due to poor performance. Led by nineteen angry store managers, the alliance declared that "from youth to old age we work devotedly on behalf of the company; therefore, we want the company to institute a sufficient severance/retirement allowance." The managers demanded ¥100 for employees with one to two years service, up to an extraordinary ¥10,000 for those with fifteen or more years of service.[5] They also called for supplementary allowances when stores were shut down. As in Tokyo the previous spring, the protesters pressured Singer by selling from inventory in hand and collecting installments as usual, but retaining all revenues in escrow. In this case the company took a harder line and fired the nineteen store managers.

This sparked a barrage of leaflets blasting the company for "the brutal violence and atrocity of the Yankee capitalists," who "caused grave problems in our constitutionally governed nation," and appealing for sympathy and support to "all brothers in the Japanese empire."[6]

A lawsuit filed by the nineteen store managers in Osaka District Court in January 1926 reflected the spirit of the dispute tactics as it also challenged Singer's property rights. The suit charged that while negotiations were taking place in December, hired ruffians forcibly removed the managers from their homes and stores (managers typically lived in the back rooms of the stores). Claiming that a branch manager was not an employee but an independent agent subject to commercial laws related to agents, the suit charged the company with invasion of property and obstruction of the right of businesses to operate, demanding payment of damages in excess of ¥10,000.[7] No reports survive as to the fate of the alliance's demands or the lawsuit, but any employee victory surely would have been invoked during the dispute of 1932–33 to legitimate the similar demands being raised. It seems certain that the managers lost their case and their jobs.

These clashes foreshadowed the tumultuous later struggle in their demands for severance pay and reform of the employment contract, in the tactic of taking over the shops, and in the combination of anti-American rhetoric with defense of a Japanese "constitutional" order. The employee complaints that would later erupt in violent dispute were both deeply held and persistent. But Singer's willingness to negotiate and consider changes for its salesmen, shown in the first of these brief disputes, would not recur.

THE STRIKE OF THE SALESMEN

The precise number of stores and employees in the Japanese empire at the time of the 1932 dispute is not known, but both sides spoke frequently in general terms of Singer's eight hundred stores and eight thousand employees. When the struggle began, Singer's Japanese headquarters was in Kobe, with central offices in Kobe, Osaka, Yokohama, and Seoul, each coordinating the activity of roughly two hundred shops. They faced a major obstacle in the changed rate of currency exchange. In November 1929 the Japanese government had returned the yen to the gold standard, an ill-timed step as the world entered an unprecedented depression. Two years later, to staunch an unprecedented outflow of bullion, the authorities were forced to abandon the gold standard. In just two weeks in mid-December, 1931, the dollar value of the yen fell from 50 to 39 cents. By August 1932, on the eve of the dispute, one yen was worth only 23 cents. In eight months, the cost of importing a Singer machine had more than doubled.[8]

The company claimed to have acted with restraint in raising the retail cash price of the most common, pedal-driven family machine by just 26 percent (from ¥185 to ¥250) between January and August of 1932. The Singer employee federation countered that the company was pricing itself out of the market. Sales indeed fell sharply under the impact first of a crash in domestic prices and wages and rising unemployment in 1930–31, and then of the yen devaluation in 1932. In both units and dollars, in both the home islands and the empire, sales in 1932 were half or less of sales in the peak year of 1929 (see figure 6).[9]

In July Singer sent its executive responsible for the Far East from New York to address the situation. From a base of operations in Yokohama's New Grand Hotel, where he would remain until February 1933, Mr. Richard McLeary determined to cut fixed costs, beginning with pay cuts, layoffs, and consolidation of the central offices. He began in August by announcing a 10 percent pay cut for all salaried staff.[10] In response, a reported eighty Singer employees at the main office in Kobe and the central office in Osaka organized a dispute group with support from the Japan Federation of Labor (Nihon Rōdō Sōdōmei), the largest union in Japan. The employees demanded the pay cuts be revoked. They held rallies and marched on Singer headquarters but had not yet gone on strike when they reached a settlement on September 11. Employees earning more than one hundred yen a month would accept the 10 percent pay cut, those earning from seventy-five to one hundred yet would take only a 5 percent cut, and those earning less kept their wages unchanged.[11] Over the next several weeks, the salaried employees at the company's Seoul central office raised similar demands and gained a similar result, presented in the Korean language press as a major victory. The majority of these employees were Korean, and their discontent was addressed as much at the Japanese supervisors as at the American company.

The company's willingness in both Kobe and Seoul to meet with representatives of the salaried workers and eventually compromise on the pay cut fueled the determination of the salesmen to raise anew their long-standing discontents. Their action began in earnest in October. Principally because of its potential to worsen U.S.-Japan relations, already tense due to American displeasure at Japanese aggression in Manchuria, this dispute was documented in unusual detail from multiple perspectives.

The Japanese press covered it extensively, the American press gave it attention when violence erupted, and the Japanese Foreign Ministry collected roughly a thousand pages of materials on the dispute, including not only its own reports but, more importantly, more than one hundred of the reports filed with the Home Ministry by prefectural governors or police officials over the course of the dispute. Together, these sources allow an unusually intimate and multisided view of an unusually long and intense labor dispute in prewar Japan.

The intensity in part reflected the personality of each side's leader. After the violence of January 1933, Richard McLeary was described to the Tokyo police by his erstwhile defenders, officials at the American Embassy, as "lacking an accommodating spirit" and bearing "partial responsibility" for allowing the situation to get to this point.[12] The principal employee leader was Yamamoto Tōsaku, by this time the manager of one of Singer's flagship stores in the Nihonbashi district of central Tokyo, near some of the grandest department stores in the land. Yamamoto was no more accommodating than McLeary, a man whose passion outran his judgment.

His hagiographic biography claims that Yamamoto began secret efforts in July 1932 to organize an employee organization to seek reform of the compensation system, ironically aided by acquaintance with other store managers fostered by a company-created Tokyo Store Managers Association. These efforts came to public attention in late September, when about twenty Kantō-area store managers and salesmen assembled at the Matsumoto Restaurant in Hibiya Park, a famous site of political meetings for three decades, to announce plans for a "Kantō Federation of the All Japan Singer Sewing Machine Employee Association" representing employees from Yokohama and Tokyo to the northeast. In short order, a parallel Kansai Federation was launched, centered in Osaka and Kobe.[13] Grand "inaugural assemblies" attended by several hundred employees followed in Tokyo on October 9 and in Osaka on October 21, while on October 11 Yamamoto Tōsaku, as chairman of a joint Employee Association, handed over their demands to Singer managers McLeary and Walker. The latter immediately announced that they "could not recognize any of the demands."[14]

In November the Employee Association set forth its claims in remarkable detail in a fifty-six-page mimeographed pamphlet. The first demand, for a rollback in the price of Singer machines, reflected the detailed business knowledge of store managers and sellers, their claim to a voice in corporate strategy, and an angry critique of capitalism. Comparing the prices of Singer parts and machines to those of competitors, the employees argued that the largest sewing machine company in the world should be able to draw on economies of scale to make a good profit even after cutting prices to levels that would allow sales and earnings to rise. The refusal to do so was the "evil custom of capitalists" who "seek the greatest profit for those at the top and leave the least for those on the bottom."[15]

The second demand, for "immediate establishment of a retirement and severance pay system" was addressed not at a generically evil capitalism but its specifically "American" form. The employees accepted the logic of capitalism by which struggling businesses dismissed people, but called for what they saw as a particular Japanese practice to deal with job loss. By the 1920s, the majority of large Japanese firms offered severance payments to employees, proportional to length of service. The government in 1926 took a small but meaningful step to codify this practice, when the Revised Factory Law (originally enacted in 1911) took effect with a new provision that employers in most cases offer at least two weeks' notice, or two weeks' severance pay, to a dismissed employee.[16] The employees maintained that "although one does not find this system in America, we think it is natural that a company doing business in Japan, opening stores in Japan, employing Japanese people, and operating in accord with Japanese law, should set up the same system of the many companies in Japan."[17]

The third demand, for "revision of the employment contract" sought change in the punitive terms of the commission system for store managers, canvassers, and installment collectors described in chapter 2. The employees objected to the several forms of "adjustments" deducted from their pay when an installment customer ran off with a machine, returned it before paying it off entirely, or paid it off in cash ahead of schedule. An additional complaint about the employment contract came from the installment collectors, who objected to the bonus system for high rates of collection. The bonuses went to their supervisors as well as themselves. To make sure they got their own bonuses, the supervisors allegedly forced collectors to advance customer payments out of pocket and present the appearance of a full collection. This placed many collectors in debt. Employees did not demand that all "adjustments" be abolished, but they asked that penalties be reduced or imposed more flexibly.

A fourth demand addressed the mandatory fidelity insurance for store managers, sellers, and collectors also described in chapter 2, which protected the company from any losses incurred by employees. The employees argued that this *fourth* layer of protection (or of distrust) was unnecessary. It stood in addition to loss-related pay deductions, employee security deposits given to Singer upon hiring, and the requirement that two guarantors share responsibility for losses. The specific demand was simply that the company issue insurance certificates to employees, who suspected that Singer pocketed the monthly premiums without actually purchasing insurance.

Appended to the lengthy pamphlet were four essays by a salesman, an installment collector named Koyama, the employees of the Tokyo Women Teachers' Division, and the Warehouse Division employees, which together evoked the emotions underlying these demands. The salesman told of being baited with promise of paid training, later revealed to be an advance repaid by deductions from his commissions. Koyama described pressure leading him and his fellow collectors to run up debts from ¥200 to ¥1,000 so as to maintain bonus-worthy records; some were led to steal machines in desperation. The warehouse workers described themselves as disposable men, easily fired in slow times and subject to inhumane, discriminatory treatment reflecting Singer's "anachronistic" attitude.[18]

The women teachers, a small minority of the disputing employees, whose presence was noted assiduously in both the press and police reports ever alert for female assertiveness, waxed particularly eloquent:

> We are given the beautiful title of *"onna kyōshi"* when hired, but immediately are told we must sell at least one machine a month, plus parts and accessories. Those who do not are let go. Even those who have worked for 10 or 20 years do not receive a single *mon* of severance pay. Hours are from 8 A.M. to 5 P.M., but to suit customers we are expected to be available after hours or on Sundays, and this is essentially unpaid work. When the company is busy, we might have 5 or 6 visits a day, but when things are quiet, we are simply laid off. And there are no year end or summer bonuses. For over 20 years the company has been treating us cruelly in this way.[19]

Through October, Singer refused to meet the employees in their capacity as association representatives, although the executives would occasionally meet them simply as employees. In one such unofficial meeting on October 25, McLeary—in his sole indication of such flexibility at any point in the dispute—told two collectors he would be willing to make some adjustment in the collection rules. They replied that a settlement of this issue alone was out of the question; all demands had to be addressed.[20]

At this impasse, the Kantō and Kansai Federations on October 25 initiated what the newspaper *Yomiuri* called a "general strike" of employees centered on the cities of Tokyo, Yokohama, Nagoya, Kyoto, Osaka, and Kobe. The great majority of participants were commission-based employees (store managers, salesmen, canvassers), joined by some women teachers and warehouse workers. The provincial shops and their sellers along with salaried staff in the central office sat uneasily between the company and their fellow employees.[21] The *Tokyo Asahi* claimed that "with the exception of Keijō [Seoul], 2000 employees in the six major cities of Tokyo, Yokohama, Nagoya, Kyoto, Osaka, and Kobe, backed by the General Federation of Labor, simultaneously went on strike."[22]

To call the action a "strike" was a misnomer. The *Yomiuri* headline the next day was more accurate: "A strange battle of endurance: collections and machines as weapons, conducting sales during the sewing machine dispute." As in the 1925 disputes, but with greater scope and for a longer time, the employees took control of Singer stores, inventory, and installment accounts. Store managers and employees continued to go door to door selling machines and collecting installments, holding the proceeds "in trust" in bank accounts controlled by the employee association. In Tokyo alone, the forty-two shops joining the action controlled an inventory of fifteen hundred machines worth more than ¥300,000 at retail prices and an even more valuable ten thousand installment customer accounts with outstanding balances of more than ¥1 million.[23]

The Kantō and Kansai Federations set up "control divisions" governed by stringent rules and procedures to ensure appropriate handling of machines and funds, including the issue of provisional receipts. Collectors would fill these out and submit them via inspectors to the Control Division, which would then pay commissions.[24] Whatever the rules, Singer of course viewed employee "control" of inventory and installments as illegal and threatening. In mid-November, the company sent postcards to all installment customers in the major cities instructing them to refuse payments to collectors and send them instead by money order directly to Singer's post office box in Yokohama.[25] Singer sent store managers a letter threatening to fire and to sue anyone who did not immediately return to the company all machines and parts, and all escrowed revenues.[26] The company followed with letters to guarantors likewise threatening legal action against the guarantor to recover these funds and machines.[27]

The employee association countered that its members were acting responsibly and legally. It also sought to draw in salaried workers as well as the shops and employees from the provinces, and a good number of the salaried workers in Tokyo, Yokohama, Kobe, and Osaka did support the strike alongside the great majority of sellers, canvassers and store managers in these major cities. But outside the major cities, only in Hokkaido and Karafuto did a significant number of inspectors, shop managers, sellers, and collectors support the dispute.[28]

Although unable to organize a majority of salaried workers or mobilize the sales force nationwide or in the colonies, the disputing workers were based in Japan's strategically and symbolically important cities. They were also encouraged by the neutral stance of the authorities, who refrained from supporting the company and in some cases indicated a degree of ethnic or nationalistically rooted sympathy with the employees. Representatives of the dispute group made the rounds of government offices and pleaded their case not only at the Home Ministry and the Tokyo police—state agencies which ordinarily dealt with labor issues—but also at the Trade Section of the Foreign Ministry and at the Army Ministry.[29] One senses the Japanese state's suspicion of Singer in a prefatory comment in the Kanagawa governor's report on the political views of A. K. Aurell, general manager of the Yokohama Central Office. Aurell was said to "require particular attention" because he was a United States military reserve officer.[30]

On December 2, McLeary, Aurell, and Walker visited the Ministry of Justice with their Japanese lawyer to defend their property rights. Noting that "among the employees are those who have not returned machines to the company but are collecting payments for them," the executives asked the government to "take strict measures."[31] The response was notably unsympathetic. The police reportedly did not view the employee actions as illegal and were not obstructing them.[32] According to the *Japan Advertiser*, officials in the Commerce Ministry, Justice Ministry, Kanagawa governor's office, and the Tokyo police infuriated Singer by refusing repeated pleas to act against the employee seizure of machines from company warehouses, collection of installment payments by the strike group, or sale of new machines. Officials deemed these tactics "merely 'part of the strike,'" and the police repeatedly told the company to "compromise with the strikers."[33] The Japanese press was as hostile to Singer as the English-language press was friendly; the *Chūgai* (forerunner of the present-day *Nikkei shinbun*) scoffed that Singer had gone "crying" to the Foreign Ministry for help.[34]

In the face of this pressure, Singer stood firm: "We are told to compromise, but a compromise would be fatal to the future of the company and is utterly impossible."[35] In mid-November, the company dismissed fifty-six salaried staff, twenty-eight warehouse employees, and twenty-four teachers who were idled by the strike.[36] Later in November Singer dismissed all employees taking part in the dispute and closed the Kobe and Osaka central offices, shifting all office and administrative work to Yokohama.[37]

As the deadlock continued into the New Year, signs of division and retreat emerged among the disputing workers. On January 12, the Kanagawa governor told the Home Ministry that the employees seemed resigned to defeat, but were trying to negotiate for at least some of their demands.[38] A major blow to their morale and solidarity came the next day. Three members of the employee association's "collection division," who had been depositing the payments they collected to an account at the First Bank in Nihonbashi, went in secret to the bank, withdrew ¥8,000, and disappeared. It turned out they had gone over to the company and were staying at lodgings provided by the Yokohama central office.[39] A jocular account of this incident appeared in the *Yomiuri* the next day, chronicling the "disappearance" of one of the three men, Katō Kantarō, a collector in his sixties. It appeared at first that Katō was taken into custody by five policemen. When he did not return home by evening, his wife went to the police, but they had not arrested or detained him. Upon investigation, they learned that Katō, who had collected ¥1,200 in installment payments in the course of the dispute, had become fearful of company retribution. In an elaborate theatrical, now exposed, he had arranged with Singer for hired enforcers to impersonate police and take him into "custody," so his fellow employees would not see him as a willing betrayer.[40]

DENOUEMENT

Incidents of violence, including brawls and window smashing, and the arrest of thirteen employees, had broken out at several points during the dispute, and the company had hired groups of toughs to protect loyalist employees and its inventory.[41] The pot boiled over on January 18, 1933. Seventy Tokyo employees, angry at the defection of the three collectors, made their way to Yokohama in small groups. Led by Yamamoto Tōsaku, they charged into the company during the lunch break when the office was almost empty.[42] There followed what the *Japan Advertiser* described as a rampage of "ruthless destruction," in which the attackers "smashed typewriters, broke tables and chairs and even tore down electric light fixtures in their wantonness. Filing cases were yanked open and valuable records were carried out, later to be burned, torn into scraps or thrown into a nearby canal. . . . The ruin of the office was complete in a matter of moments."[43] Company strongmen and employees rushed back from lunch to defend the premises. The ensuing battle sent nine men to the hospital with serious injuries. Police arrested more than forty employees.

The attack marked a turning point. Singer executives in Yokohama threatened that they would sooner quit the Japanese business entirely than make concessions. Singer's head office in New York backed this up with its first semipublic statement on the dispute, an order to the Japanese office, strategically leaked to the press and police, to make no concessions and to refuse to meet or recognize the employee union. The New York office also announced willingness to rehire strikers on an individual basis if they quit the dispute group.[44] In Tokyo, Ambassador Joseph Grew complained to the Foreign Ministry, while Secretary of State Stimson raised the matter with the Japanese ambassador in Washington.

Against this background, the Kanagawa prefectural police, with jurisdiction over Yokohama, mediated a settlement.[45] Near midnight on February 9, after a final day of melodrama and negotiation, the dispute group abandoned virtually all its demands but won some slight face-saving concessions. The company agreed to make commissions and other terms of employment the same for employees in all six major cities, in essence raising these levels to those of the Tokyo employees. It also pledged to issue certificates for the insurance premiums, the one thing it had offered to do when the initial demands were submitted in October. Excluding those who had "engaged in criminal acts," it agreed to rehire the fired workers at company discretion based upon interviews. A side agreement clarified that to be considered for rehiring, salesmen and collectors must hand over all funds collected during the strike within ten days; if they did so, the company would not pursue legal action. The point which stuck in Singer's throat and nearly derailed the settlement literally at the eleventh hour—but was standard practice in labor disputes at the time in Japan—was agreement to pay a large lump sum (¥7,500) to the dispute group to cover its legal costs and provide financial help to jailed employees and their families.[46]

Not surprisingly, bitter feelings persisted on both sides for some time. Singer suffered—or accepted—a considerable exodus of employees. Attrition in Tokyo was particularly high. The company rehired only 6 of 37 warehouse workers, 11 of 45 women teachers (in part because it permanently shut down its school in central Tokyo), and 42 of 76 installment collectors. Singer rehired "only a few" of the two hundred Tokyo salesmen who took part in the strike, while the exact fate of the 42 store managers in the city who took part remained unresolved at the time the final police report was filed in late March. Thirteen managers had been dismissed for breaking the terms of their employment contract, while sixteen had been judged "relatively loyal" and apparently rehired. Another thirteen were still arguing with Singer over the terms of their return, reportedly while exploring other options, such as importing sewing machines from the Soviet Union, selling Japanese-made machines, or dealing in used machines.[47]

"NATIONAL CUSTOMS" VERSUS "UNIVERSAL" SYSTEM

In some respects the Singer employees, including even two hundred store managers and a number of regional inspectors, were attacking capitalism in general. They did not see exchange-rate fluctuations as a legitimate reason to raise prices. They blasted as unfair the adjustments that linked their commissions to actual rather than pledged payments by customers. They squarely challenged the company's property rights by seizing inventory and installment accounts. Such views and actions reflected a moral economy of the employees that might well have been mobilized against a Japanese company with similar policies. And had the opponent been a Japanese company, the government—as it did in the face of production takeovers in 1945–46—might well have stood with Japan's capitalists against their workers.

But this incident was more than a generic dispute between capital and labor. The rage of the disputers—and the sympathy of the state—were obviously reinforced by a comparative understanding of economic life that directed hostility toward a specifically "American" capitalism. In private negotiating sessions, dispute group representatives scorned the "traitorous attitude" of high-level Japanese managers working with McLeary and Aurell. Such Japanese "threaten their subordinates and suppress the national feelings [of employees]."[48] Shrine visits were not part of the repertoire of labor dispute actions in this era, but in this case fifty dispute group members made a point of offering prayers for success at both the Yasukuni shrine to Japan's war dead and the Meiji shrine to the revered grandfather of the reigning emperor.[49] Leaflets and public statements by the dispute group condemned Singer as a "Yankee capitalist" company, making the workers doubly victims of ethnic and class oppression.[50] As the employees grew more desperate, their rhetoric became more virulent: Singer "spoke contemptuously toward the great Yamato race" (*minzoku*); the Japanese managers who supported the company were "public enemies, traitors, and betrayers, insects on the loyal body of the nation. Such people should be expunged from our imperial nation."[51]

The racial and ethnic references of the Singer employees occasionally revealed how uneasy their position was as Japanese standing between a West seen to oppress them and an Asia—and a world of people of color—seen as beneath them. At a meeting on October 22, a dispute-group representative told McLeary and Aurell, "change your attitude, which sees us as Negroes. Don't you understand the position of Japan in the world today?"[52] Nakai Ryūnosuke, the chief clerk in the Yokohama Central Office, was a leader among forty-one salaried employees, at one point reported to have joined, and then almost immediately quit, the strike group. A group of forty-two outraged dispute-group members forced their way into the Yokohama office to denounced Nakai as "an American spy who, although Japanese, has gone foreign." They urged the office employees to rejoin the employee federation, apparently with some effect. Nakai later stated that, "although we do not wish to join the federation, as Japanese we unavoidably feel pressed to join it."[53] For their part, Korean employees in the Seoul office dispute of September 1932 mirrored this triangular racial politics, arguing that "the way we have been treated is unacceptable. Even more than the fact that a Westerner treats us in this manner, it is the actions of the [Japanese] managers working under him that anger us."[54]

The rhetorical and conceptual tissue connecting ethnic or racial pride to scorn for American capitalism was the notion of Japan's "national customs" (*kokujō*). Foreign managers were described as "hairy barbarians who do not sufficiently understand our national customs." As they raised the flag of "national customs," the Singer employees described the substance of the customary Japanese practices they expected employers to follow. They called for severance or retirement pay linked to seniority, well aware that this was increasingly common for both white-collar and production workers at large Japanese corporations and had been encoded to a modest degree in law. In making this demand, they claimed repeatedly "that insofar as [Singer] operates in Japan under the protection of Japanese law, it should follow Japanese custom."[55] And indeed, three years later Japanese law itself came to follow this path of custom more fully; in 1936, the Imperial Diet passed a law requiring all businesses with more than fifty employees to offer severance pay linked to years on the job. For long-serving employees, this amounted to a modest pension. The debate over this law was animated by the idea that corporations and the state had responsibility to design "appropriately" Japanese social policy.[56]

While still on the job, the disputing employees expected stable wages and job security. Their claims reflected an idea, gaining force by the 1920s among Japanese working people, that, in modern capitalism, employees—women as well as men—held a "right to benevolence," including rights to their jobs in all but the most dire of business circumstances.[57] Singer's women teachers expressed fury that "even those who have worked for 10 or 20 years" were let go "without a single *mon* of severance pay" if they failed to meet the sales quotas added to their teaching duties.[58] The male canvassers charged that Singer's complex array of security deposits, insurance, and guarantors was intentionally structured to promote dismissals; it offered perverse incentives to fire salesmen or collectors whose customers were in default, since "all bad debts are immediately collected. In this system, an employee who performs poorly [and is fired] actually profits the company."[59]

The calls for severance pay and criticism of compensation policies and work contracts gave a substantive grounding to the rhetoric of the employees who condemned Singer as anti-Japanese for not respecting "national customs." Against such logic, Richard McLeary and his managers in Japan, with the support of the New York headquarters, justified their inflexible stance with a consistent chorus of claims for the universality of the Singer system over time and across the globe. In a lengthy English-language document submitted to the Foreign Ministry and distributed in Japanese translation to employees and the press, responding to the equally long statement of grievances by the employees, the company proclaimed that "the proof that the Singer system is a good system is the fact that it is universal throughout the world." Further, "every person who enters our organization comes in knowing that there is no retirement allowance system and this is not only true in Japan but in every other country in the world. Referring to the question of making alterations in the present commission contract, . . . they [the disputing employees] entirely disregard the broad principles of the various clauses in the contract based on which our business has been maintained so successfully for so many years." The occasional abuses of the system by particular managers "cannot be taken as reasons for condemning the whole system and the Singer service contract now in force must stand in its entirety if the Company is to continue the present system of business in Japan."[60]

These claims were certainly forceful, and they reflected a long-standing consistent stance. Already in the 1890s, Singer's president, Frederick Bourne, praised the "simplicity of the rules and regulations" of Singer's selling system for allowing "thousands of nationals of all cultures, many of whom were previously unacquainted with American business procedures," to effectively put it into practice.[61] But these claims are nonetheless puzzling in view of Singer's willingness elsewhere to improvise in response to local conditions and compromise when employees objected to its policies. In a number of countries in the late nineteenth and early twentieth centuries, including Britain and Russia, Singer shifted from separate cadres of canvassers and collectors to a system where salesmen both sold machines and collected installments.[62] In India in 1901–2, Singer's vice president told the manager of operations not to worry about adopting the company's sales practices "word for word and line by line."[63] In 1906–8 in Russia, Singer's managers demonstrated "structural flexibility and procedural adjustment" in the face of perceived employee resentment by changing strictures applied to guarantors, the fidelity insurance system, and rules of the employment contract. Singer's overriding goal, in the words of Alfred Flohr, the successful German general manager of the Russian operation, was to "develop the business to as great an extent . . . as possible," even if this meant adjusting Singer's famed system.[64] The company's sprawling operation in imperial Russia faced occasional strikes, most notably in Warsaw between 1905 and 1909. In this case, Flohr acknowledged "serious inequities in the structure of commissions," and he agreed to make changes in the "commissions and office procedures."[65] And even in Japan, we have seen that in the first dispute of 1925, Singer promised employees to seriously "consider" a system of retirement and severance pay, something they would not even hint at considering seven years later.

The presence of organized employees with connections to a national labor federation was certainly a sticking point for the company in the fall of 1932, and it goes some way toward explaining Singer's intransigence: "Any individual member of the organization [who] has suggestions, complaints or petitions to present to the Company . . . is at liberty to do so in his personal capacity but not as representing any group."[66] Yet Singer had been willing to negotiate with organized employees in Tokyo and Osaka in 1925, as well as in Seoul and Osaka in the first round of disputes in the summer of 1932. The Osaka employees that summer had affiliated with the Japan Federation of Labor, and Singer negotiated not only with the federation but with the employee association in Seoul. And in the end, in February 1933, Singer did allow its lawyer to work out a final settlement through negotiations with the federation's Matsuoka Komakichi. Anti-unionism was certainly a factor reinforcing Singer's rigid approach, but it was not the whole story.

Neither was a commitment to a particular ceiling for compensation the main source of Singer's uncompromising stance. After imposing pay cuts in the summer of 1932, the company negotiated with employees and modified its position. McLeary also on one occasion—seeking, it would seem, to drive a wedge between the two groups—hinted at willingness to slightly modify the rules for collectors but not for canvassers. What Singer absolutely refused to countenance was any change in its "universal" system that would provide any sort of fixed element—including severance pay—to the compensation of its sales force.

The available evidence offers no clear explanation for this intransigence. Perhaps the depth of the depression made Singer cling to familiar practices more fiercely than ever. But I suspect that continued decades of global domination had eroded an earlier willingness to adjust what were increasingly long-proven practices. Whatever the reason, Singer's confidence in its system had hardened into rigid faith. In the aftermath of this struggle, an alternative faith would emerge: the disputing employees joined fledgling domestic competitors to clone Singer machines but modify the Singer system.

FROM FIGHTING YANKEE CAPITALISM TO ADAPTING THE SINGER SYSTEM

Shortly after the settlement, Richard McLeary is said to have told his fiery opponent, Yamamoto Tōsaku, that "we're done for in Japan" and to have explained to foreign residents in Yokohama that "even if a foreign company wins this sort of labor dispute, it will fail in the long run."[67] These statements, cited by Yamamoto's sometimes hyperbolic biographer, might be apocryphal. But Singer indeed lost market share through the rest of the 1930s. The company's decline in Japan is made clear by comparing its sales to that of its new domestic competitors (see table). The slow recovery of Singer in the Japanese home islands compared to its rapid expansion in Korea, where it suffered no prolonged strike, suggests that the dispute had a significant impact on business. In the home islands, sales after the strike never recovered to the level of the 1920s, which ranged from forty to sixty thousand machines each year. Yet Singer sales in Korea, which had peaked at just under twelve thousand machines in 1929, rose to new heights in both absolute and relative terms: total sales in 1937 were double those of 1929. Korea's share of all sales in the Japanese empire reached more than one-third by 1936 and 1937. Singer's anemic sales in the home islands did not reflect a lack of demand. On the contrary, in these very years desire for sewing machines soared as growing numbers of young working women and older adults turned to Western dress. But

HOUSEHOLD MACHINE SALES BY SINGER AND DOMESTIC MANUFACTURERS
IN JAPAN AND KOREA, 1933–1940

Year	Singer: Japan Sales	Singer: Korea Sales	Domestic Makers: Japan Sales	Domestic Makers: Korea Sales
1933	13,820	4,742	5,000	—
1934	24,259	8,591	12,000	—
1935	31,085	16,810	12,301	—
1936	34,014	22,628	40,924	3,720
1937	40,694	25,246	53,133	7,240
1938	13,024	8,106	104,204	6,069
1939	4,450	2,824	132,997	10,810
1940	971	370	154,402	11,201

SOURCE: Singer Annual World Results, 1933–40, Singer Sewing Machine Archives at Wisconsin State Historical Society and Janome mishin shashi hensan iinkai, ed., *Janome mishin sōgyō 50 nenshi* (Tokiwa shoin, 1971), p. 809.

the domestic producers such as Mitsubishi, Pine-Janome, and Brothers captured most of this growing market; a negligible presence as late as 1933, they together exceeded Singer sales in Japan by 1936.

In 1937–38, this competitive field was transformed. An import-export control law in October 1937, followed by a new foreign exchange law in 1938, gave the government full control of imports and of access to foreign currency. From late 1937, it allowed sewing machine imports only for military use—production of uniforms or boots—and the government directed all foreign exchange toward the import of strategic resources such as oil. Denied access to dollars or pounds, Singer was unable to import new machines or parts for any purpose. Sales fell to near zero after 1938. The company gradually sold off its remaining inventory, closed most of its shops, and released most employees. Japanese producers proved quite ready to step forward to replace Singer; as war intensified on the Asian continent through 1940, domestic producers in total were selling double the number of sewing machines that Singer had managed to sell at its peak in the 1920s.

For some time Japanese machinists had been creating the foundation for a domestic sewing machine industry in hundreds of small businesses that repaired sewing machines, imported and serviced used machines, produced machine parts, and in a few cases manufactured finished machines. One sign of this growing potential is the strange story of Wilansky-Goldberg and Company. In the final days of the Singer dispute, a rumor reached Osaka prefecture's governor that one M. Goldberg of Wilansky-

Goldberg, a seller of used sewing machines with a home base on West Twenty-fifth Street in New York City, had arrived in Japan to support the strikers by supplying used machines, which they sold in place of Singer machines. Further investigation revealed this report to be misleading. Goldberg, the amended story went, was simply a "typical Jew" seeking profit where he could find it. His company had opened an Osaka office in 1930 and was doing about ¥30,000 of business a month by the end of 1932. The Singer employee federation had indeed approached this office about selling their machines during the dispute, but no significant business had resulted. Although Wilansky-Goldberg did import some used machines for resale, its main business in Japan was the *export* of Japanese-made sewing machine parts. With an eye on the lower value of the yen, Goldberg had recently come to Japan to grow the export side of the business.[68]

Building on the manufacturing base revealed by this episode, by the late 1920s and early 1930s a few domestic companies were beginning to produce and sell small numbers of their own machines. Among the most promising were Mitsubishi, Yasui (later, Brothers), and Pine (later split into Pine and Janome). Mitsubishi Electric had purchased the patent of a Japanese machinist and began to produce sewing machines in 1931. The Yasui patriarch in Nagoya had been repairing sewing machines since 1905, and his two sons (source of the name "Brothers") had been making machine parts since 1925. With exquisite timing, they produced their first sewing machine in late 1932 and opened a factory the next year.[69] The Pine Company, founded in 1921, took its name from the fact that its two founding partners each had the character for "Pine" (*matsu*) in his name. Although it took years of trial and error to successfully produce small batches of machines in 1929, and several more years to figure out how to mass produce its machines effectively, Pine ranked among the most creative Japanese firms in sales and marketing.[70] Along the way, one of the founders left the company in a dispute with his partner, taking the "Pine" name with him. It was the company he left behind that truly thrived with its famous Janome model. This eventually became the company name.

These companies were emerging before the Singer dispute. But the prolonged incident offered more than temporary relief from their mammoth competitor. It stirred patriotic "buy Japanese" sentiments and led Singer employees to turn their expertise and energy against the company. During the strike, in addition to approaching Goldberg for used machines to sell, Singer salesmen sold domestic machine parts, reportedly earning

FIGURE 15. Compare this photograph of the first Nippon Sewing Machine, sold by Yamamoto Tōsaku on behalf of the fledgling Brothers Sewing Machine Company, to that of Singer's family machine of the same era (figure 16). It is clear that Brothers had produced an exact copy of the Singer design, including decals and the typeface used for the company name. (Nature and Science Museum, Tokyo University of Agriculture and Technology)

a healthy average of two- to three-yen daily profit per person.[71] In late December, an organization calling itself the Singer Employee Dispute Support Group passed out leaflets in Osaka calling for a boycott of the company's products for the duration of the dispute; boycotts were reported elsewhere as well.[72] Yamamoto Tōsaku was the most famous Singer employee to turn to the domestic industry after the dispute. The Yasui brothers, based in Nagoya, engaged him to expand their reach. He initially marketed their machines from Tokyo to the north of Japan under the name of Nippon Mishin.[73] This machine (figure 15) was a clone of the Singer family-type machine (figure 16), and Yamamoto brought his Singer sales know-how to the task as well. He helped Brothers emerge by the end of the decade as one of Japan's most successful producers.[74]

Certainly firms such as Brothers, Mitsubishi, and Pine-Janome eventually would have begun to compete more successfully with Singer. But in 1932 they were as yet small and fragile; if not for the opening provided by the dispute, they surely would have taken longer to build momentum. An existing determination to build a domestic sewing machine industry to compete with Singer was heightened by the patriotic or anti-

FIGURE 16. A Singer family-type machine sold widely in Japan and around the world in the late 1920s and early 1930s. (Nature and Science Museum, Tokyo University of Agriculture and Technology)

American sentiments unleashed by this labor struggle. This conjuncture helps explain the determination of these domestic producers not only to adopt but to creatively adapt central elements of the Singer selling system.

Janome, like Brothers, cloned Singer machines and adopted many of its sales practices, but Janome also loudly and proudly adjusted the Singer system. It was walking a rhetorical tightrope; the company wanted to defeat Singer and position its good as part of a Japanese-styled modern life, but Janome understood the deep connection of its product to the American brand. In the mid-1930s, it advertised its machines as "the same . . . as the Singer model 15" but "half the price" (figure 17). It set up a selling network of shops and shop managers identical to Singer's, with salesmen using customer cards to divide territory into "blocks." In what the company's in-house historian called a "dark" legacy of "Singer's punitive thinking," Janome also made shop managers completely responsible for any losses incurred through default on installment sales.[75]

But Pine-Janome did make one major change in the compensation structure that should be understood as a strategic response to the complaints of Singer salesmen—some of them now in their employ. It abandoned straight commission payments in favor of fixed pay plus commis-

FIGURE 17. This ad for Pine Sewing Machine's Janome model appeared in *Asahi shinbun* (2/26/1934, p. 5). It makes clear that Pine-Janome was simultaneously seeking to draw on the prestige of its American competitor and put itself forward proudly as Japanese. The lettering next to the box in the center emphatically reads, "The same large Janome type as the Singer model 15." The line on the far right side reads, "Half the price of the foreign good." The phrase "our country's product" (*kokusan*) precedes the model name, "Janome," and the large characters in the centered box boast, "Government Assistance." (Janome Sewing Machine Corporation and *Asahi shinbun*)

sion. In the mid-1930s, the company paid salesmen in Tokyo, Yokohama, and Osaka fixed monthly wages of fifteen yen, and those elsewhere a wage of ten yen, plus commission. This step was not without detractors. A Janome shop manager in Tokyo's Asakusa district claimed (in retrospect) that "fifteen yen a month guaranteed a minimum livelihood, and many salesmen were attracted by that and lacked strong motivation to work hard. Ten sales contracts per month was the norm, and this wasn't so hard to achieve, but most sellers were happy with four or five a month, which earned them about 40 yen." The company drew these salesmen from two sources: "a cadre of 'converts' from Singer" as well as "those who had lost their jobs in the depression or seen their small businesses fail."[76] The company put significant effort into training this sales force, writing a detailed manual entitled, in English, "Our Salesmanship." This was required reading for all employees. It offered hints on such topics as how to convince reluctant husbands to spend money on an object used by women—the machine would in fact benefit the entire family's wardrobe and budget.[77] The company grew steadily, selling six thousand machines in 1937, about nine thousand in 1938 and fourteen thousand in 1939, including some exported to Asia and to Japanese emigrants to Latin America.[78]

Like Singer, Janome made the majority of these sales through installment plans, and its second significant adaptation of the Singer system was a new credit plan, aimed at less affluent customers. Along with the nationalistic leaders of other domestic producers, Pine's first president, Ose Yosaku, believed Singer had alienated potential buyers and developed a reputation as arrogant because of credit requirements that were too stringent for lower-income buyers. Ose set his sights on what he called the "plebian class" with a modified installment plan that he believed fit "the Japanese situation." Customers who could not afford a down payment could "reserve" their machines by paying five yen a month for six months *before* taking possession. By that time, Pine had its down payment in hand, and the customer had three choices: pay the remaining cost in cash, convert the balance to ordinary installments with interest payments, or continue the layaway at a discount against the cash price without taking the machine home until fully paid. This third choice in effect earned interest on the money deposited with Pine over the course of two years.

Pine first offered this layaway-installment scheme in December 1930, calling it "monthly reservation payments." The Ministry of Finance objected that Pine was violating banking laws by acting as a savings bank. But Ose reportedly overcame these objections with an appeal to national interest, arguing that it was more important to nurture a domestic sewing machine industry than to worry about legal niceties. In the words of Shimada Takuya, a talented salesman and later company president, Pine was cultivating "the bottom of the social pyramid," the "ordinary masses," such as "the wife in a back-alley tenement who carries her baby half-asleep on her back to go shopping."[79] Pine-Janome also touted this plan as especially suited to a family with infant daughters, because the parents could start paying at birth; when the youngster was ready to learn sewing, the family would own their machine.

This creative financing strategy offered the maker a cash-flow advantage; the prepayments essentially financed the subsequent installment credit. Prepaid orders also helped the company to calibrate production schedules to future demand. And a customer sufficiently disciplined to make regular prepayments—in essence a saver, not a spender—was likely to continue meeting the installment obligations. However creative and effective they were, such plans were not unique to Japan. In the United States, Ford and Chevrolet had offered them in the early 1920s.[80] But it was significant that Pine described its "reservation payments" as an adaptation of traditional Japanese collective mutual aid known as *kō* or *mujin*. Earlier in the century, these had been the inspiration for a loan plan of the Japan Real Estate Bank, which in turn had inspired President Ose to develop his plan.[81] Whether Ose also knew of the American precedents is beside the point; by describing his plan as uniquely "Japanese," he gave it a powerful appeal in the nationalistic context of the times. The Pine-Janome managers understood and presented its selling system as better suited than Singer's to the expectations and incomes of Japanese sellers and customers alike.

State support was another factor behind the rise of a domestic capacity to manufacture sewing machines. Certainly the laws restricting imports and access to foreign exchange in 1937–38 erected a solid wall of protection that lasted far into the postwar era. And prior to this, the state acted as cheerleader and occasionally financial supporter of domestic producers. Pine's advertisements from the mid-1930s, as in figure 17, boasted in large print of "Government Assistance." This referred both to cash subsidies and to certificates and trophies awarded to high-quality domestic products by the minister of commerce.[82] Such endorsement was part of a "buy Japanese" movement that began in 1928 when the ministry announced a list of 118 "encouraged domestic products." The movement's peak came in 1930. The ministry distributed hundreds of thousands of "buy Japanese" pamphlets and expanded the list of "encouraged" goods to 312.[83] The impact of such official endorsements of quality and calls to buy national is difficult to assess. It surely mattered to some potential customers that Janome machines were cheaper than comparable Singer models. But other buyers might have worried that even a significantly less costly Japanese machine would not be of a quality worth purchasing. For them, the state's stamp of approval might have tipped the balance.

As competition mounted, Singer finally showed some flexibility by allowing non-exclusive sales representation. This is evident from the 1935 commemorative photo (figure 18) in which the large banner above the show window announces that this once exclusively Singer store now sold both Singer and Pine-Janome. But this appears to be a defensive move, too little, too late. It is fair to conclude that the pace and extent of Singer's decline were accelerated by its stubborn defense of its global system in the early 1930s. A more supple response to employee complaints would have slowed the outflow of talent to competitors. Especially as Western dress grew in popularity among adult women, willingness to ease terms of credit would have reached a wider swath of these customers. As such steps slowed the emergence of domestic challengers, Singer might still have been flourishing in Japan in 1937–38 and conceivably would have found its machines valued as strategic assets by the military and treated as a special case under the new foreign exchange regime.[84]

Of course, one must not drive this counterfactual case too far. As it did with GM or Ford, the government as it mobilized for war eventually would have pressured even a thriving Singer offering a relatively unique strategic good to join forces with one or another Japanese competitor.

FIGURE 18. This photograph of the Mochizuki Sewing Machine store in Mito City, taken in October 1935, commemorates the awarding of the Minister of Commerce trophy and plaque to the Pine-Janome Company in 1935 and kicks off the store's "special sale of Pine and Singer" (as announced in the banner above the seated employees). The trophy is held in the lap of the second man from the left, next to a framed citation from the ministry. It is noteworthy that Singer is now allowing its formerly exclusive stores (this store is at the same location, under the same management, as the shop in figure 4) to sell competing brands as well. (Mochizuki Yoshimasa)

The company surely would have been on the defensive before 1941.[85] In the actual case, Singer had been losing ground for several years before its virtual ouster in 1937–38, except in Korea, where it did not suffer from the dispute and did not find itself on the wrong side of ethnic resentments. The reasons for its decline included both the self-inflicted loss of momentum and talent in 1932–33 and the innovations of Janome and others, beginning shortly before that time.

• • •

For salesmen to lead a labor dispute is unusual. One expects sellers to be among the truest believers in capitalism. They were the people who in Walter Friedman's telling of the American story "pushed [this economic system] with its vices and virtues, into every corner of the globe."[86] Yet the Singer sewing machine salesmen were neither missionaries of capitalism nor uncritical in their faith. They led one of the longest labor disputes in Japan's prewar history—the only dispute with such a wide geographic reach; the strikers mobilized strong support across six major cities in the home islands and in the colonial outpost of Korea, including among Korean workers. The intensity of this action was certainly due in significant measure to a rising tide of ethnic nationalism; this was a moment when Japan's military-led expansion in Manchuria incited nationalism at home in the face of sharp criticism abroad. But the strike also reflected the resistance of the employees to the logic of market capitalism, pejoratively identified as peculiarly "American." The Singer employees were among many who sought treatment that they identified as "appropriately" Japanese and that they believed would offer them greater security and respect. The dispute was thus part of a stream of activism that gradually put in place elements that were coming to be defined as a particularly "Japanese" employment system.

For a company to resist such a dispute is not unusual but was in retrospect unwise. Singer's position in Japan would have weakened eventually as domestic competitors emerged. But a stubborn defense of a universal system hastened this day; it gave impetus and opportunity to Japanese manufacturers and sellers who would appropriate much of the Singer method even as they styled themselves proudly as Japanese. These producers were self-consciously localizing market capitalism as they reinforced concepts of employment and improvised selling practices understood to accord with "national customs." In so doing, they met the expectations of sellers for "appropriate" treatment and found ways to reach a broader customer base. Such firms gradually learned to produce machines of comparable quality to Singer's, through what would be described years after World War II as a Japanese-style production system, and they built the Singer sales system into what would be seen as a Japanese mode of retail sales.

PART TWO

Sewing Modernity
in War and Peace

CHAPTER 5

War Machines at Home

Rates of sewing machine ownership more than doubled over the 1930s, reaching nearly one in ten households by the decade's end. Demand remained strong into the early 1940s. The enduring desire for this good and for the Western-style dress that it fabricated reflected the force of a modern spirit that for several decades had embraced technologies and lifeways identified with the West in general and America in particular. Unfolding from the late 1930s, a drive to reform both men's and women's dress reflected the ongoing and related search for a Japanese-inflected modern life. As both object of desire and tool for reform, the sewing machine tracks the story of an expanding consumer society in an era of wartime modernity.

In one view, *wartime modernity* is an oxymoron, for Japan of the 1930s veered sharply from the cosmopolitan and internationalist 1920s into a "dark valley" of militarism and antimodern struggle. Focusing in particular on daily life, the cultural historian Minami Hiroshi and his colleagues argue that an era dramatically repudiating the modernism of the 1920s began in the mid-1930s, when the "establishment controlled and managed the behavior of the masses."[1] This change included a "break in the Japanese evolution toward Western dress" because of the militaristic imposition of uniform dress.[2] The war is said to have interrupted both the spread of household sewing machines and of women's Western dress, as women all over the land were forced to adopt the traditional agricultural work outfit known as *monpe*.

Certainly there is a foundation for such a perspective. Authoritarian control over political life expanded steadily in the 1930s. Efforts to regulate daily life intensified, especially in 1937 after a full-scale war began in China. And women did turn to *monpe* trousers en masse. But recent scholarship has shifted the argument on both causation and chronology by arguing that "the maturation of modern institutions and not their stunting led to the burst of expansionism of the 1930s," and that "a Japanese modern mass-culture" indeed "continued into the Pacific War because the consumer-subjects of the patriarchal Japanese family-state did not want to let go of the modern."[3] This new approach has centered primarily on the public face of culture and economic life, such as department stores and cinema. The story of the household *mishin* and its uses takes it into the back alleys, the homes, and the wardrobes of working- and middle-class families, allowing us to recover a sense of the ordinary agency of women as both subjects and consumers.

WARTIME MODERNITY IN DAILY LIFE

By the end of the 1920s, the Singer sewing machine and the radio—one product decidedly branded, the other relatively generic—had become the two most notable products of machine civilization to find a modest place inside middle-class homes in Japan and its empire. It is important to recognize both the material limits of this foothold and its far greater reach in people's imagination. Neither item was owned by more than 5 percent of households in 1930. Other than electric fans and irons, none of the goods spreading rapidly in the United States at this time, such as refrigerators, stoves, washing machines, or phonographs, found even this level of use, and only a small minority of Japanese could afford to live in the modern "cultural" homes built by railway and real estate developers.

Against this background, the 1930s was a time of both mobilizing for war and deepening of modernity. The level of media saturation increased sharply thanks to new technology, increased prosperity and purchasing power, and the efforts of state and private actors to mobilize for war and chronicle it. Radio-subscriber households soared tenfold from 650,000 in 1929 to 6.6 million by 1941, rising to 7.5 million in 1944. By early 1940 more than 56 percent of the nation's urban households subscribed to the national radio broadcasting service (NHK); in the countryside, one in five households were subscribers.[4] Radio broadcasts of Western sewing lessons began in 1926, NHK's second year of operation.[5] The radio was both a widely owned good and a primary means of bringing additional goods, information and dreams—whether of economy, empire, and conquest; of education and culture; or of sports and amusement—into the lives of millions.

The surge in radio subscriptions resulted in significant part from intense media competition to cover and to glorify Japan's new imperialism of the 1930s. Beginning with the takeover of Manchuria in 1931, newspapers such as *Asahi* and *Mainichi* competed with each other and with NHK to be first with the top stories; they used airplanes to send reporters and film to and from the continent, and they turned to newsreels and movie theaters, as well as newsprint, to reach a fast-growing mass audience. Their competition produced a more homogeneous national news market in which local outlets lost out to the big three of *Asahi, Mainichi,* and NHK, and it blurred the line between news and entertainment as war newsreels were shown in tandem with Hollywood movies.[6]

The pursuit of war coexisted throughout the decade not only with entertainment but with the pursuit of leisure and fashion. Even as tension rose between the United States and Japan over the latter's aggressive foreign policy, the American "national pastime" of baseball reached new heights of popularity, sparked by a famous exhibition tour by Babe Ruth and other stars in 1934. Commercial sporting spectacles coincided with the new services and habits of personal decoration, especially for women, part of a quickly changing urban landscape centered on the middle-class consumer. Hair salons spread throughout Japanese cities, offering permanents to thousands of middle-class women; by 1939 these shops numbered eight hundred and fifty in Tokyo alone. Of even greater interest for women, and fascinating for the culture critics and the male-dominated mass media, the 1930s—more than the "modern" 1920s, with their campaign to reform daily life and debates over the merits of Western and Japanese dress—saw a significant acceleration of the shift toward Western dress for women, on the grounds of both fashion and purported rationality.

The most important interpreter of this shift during the era of wartime modernity was Kon Wajirō. Born in 1888, he trained in Tokyo as an architect and was then appointed to the faculty at Waseda University. From the 1920s through the 1960s, Kon offered a brilliant ethnography of modern daily life through numerous essays in both the popular press and academic outlets. His trademark was the idiosyncratic social survey, cleverly illustrated with graphs. In one of his earliest efforts—the surveys carried out in Tokyo's Ginza district in May 1925, mentioned in chapter 3—Kon had found that 67 percent of men but just 1 percent of women wore Western dress.[7] He reprised this investigation eight years later, at

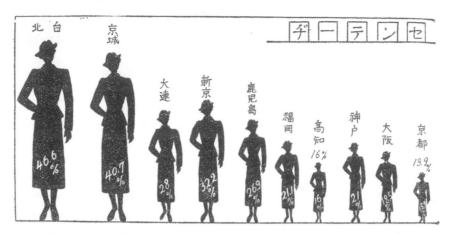

FIGURE 19. Titled "The Percentage Wearing Western Dress" in various cities, this graph is characteristic of Kon Wajirō's eye-catching presentations. It uses the hand-drawn figures of women to represent the spread and relative frequency of Western dress in cities. The figure on the far right represents the average across all locations. The two tallest figures on the left are from the colonial capitals of Taiwan and Korea. Undertaken in May, the survey was published in the June 1, 1937, issue of *Fujin no tomo*, pp. 93. (*Fujin no tomo* publishing company and Ochanomizu library)

3 P.M. on February 25, 1933. With their precise time-stamping buttressing a sense of social scientific precision, his observers found that 19 percent of 462 females wore Western dress, at first glance a huge change from 1925. But children or students accounted for the great majority. Subtracting them, Kon counted only 2.6 percent of adult women in Western clothes. Kon saw his results as a corrective to the common wisdom that women were increasingly turning to Western dress. He argued that this erroneous impression arose because of disproportionate attention given to those in the forefront of change.[8] Two years later, in a good example of this same exaggeration, Kon reported a "remarkable increase" in the sale of stylebooks for women's Western dress; women now "swarmed" the city streets looking as if they had "just stepped off the page of stylebook drawings." He found something "fresh" about such Western dress in Japan compared to the "past-its-prime" fashion of Japanese dress.[9]

In the following years, Western dress continued to spread among city women. Two months before Japan launched full-scale war in China, on May 1 and 2, 1937, from 3 to 4 P.M., a large team of volunteers mobilized by *The Lady's Companion* helped Kon with his most ambitious survey yet. This remarkable snapshot of exactly 26,002 females in eighteen cities found that 26 percent wore Western dress. Kon's playful graph of

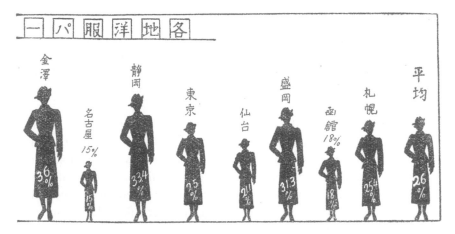

his results (figure 19) showed that Tokyo stood just below the empire-wide average, while Taipei and Seoul, as home to the most Westernized women of all, stood tall. A subset of adults not categorized in earlier surveys, the much-discussed "working woman" (*shokugyō fujin*), accounted for nearly one-fifth of those in Western garb. If we apply Kon's discount for the high rate of children and students in Western clothes in 1933 to the data he presents for 1937, about 6 percent of adult women, including the "working women," were now strolling in Western dress, more than double the 1933 proportion.[10] Summertime was particularly conducive to lightweight Western wear; according to a different survey, nine in ten women in their twenties and one-fifth of those in their forties had adopted Western summer dress by this time, prominently including the *appappa*.[11] This steady increase in women who were wearing one-piece dresses, skirts, and blouses was accompanied by a persistently anxious discourse over dress, war, and national identity, examined in detail below. For the moment, it suffices to note that up to and into the years of war, Western dress modes were broadening their appeal.

Simultaneously, and related, both sewing schools and dress shops surged in popularity. They built upon a base of Western dressmaking schools established in the 1920s. Some had been founded by men who had formerly sold sewing machines or operated garment shops. Among the most successful was the Women's Academy for Cultural Sewing (Bunka saihō jogakuin, introduced in chapter 3), which enrolled more than one thousand students by 1933. Just three years later, it boasted thirty-five hundred students and had changed its name to the more broadly cast Academy for Cultural Dress (Bunka fukusō gakuin). In addition, several women who had learned dressmaking in the United States or in Europe opened schools in the 1920s, offering access to the latest techniques as a selling point. Among the most famous was Sugino Masako, whose Dressmaker Women's Academy (colloquially known as Doreme, contracted from the Japanese pronunciation of "dressmaker") grew to enroll twelve hundred students by 1936.[12]

These schools offered a bridal course of training for domestic sewing as well as lessons for women seeking vocational skills to support themselves or their families. Bunka fukusō and Doreme each devised proprietary pedagogy that targeted home sewers more than professional dressmakers. These and other schools formed mutually beneficial alliances with women's magazines to reach housewives in particular, placing ads for their classes in multipage supplements.[13] These expanded significantly over the course of the decade. In the late 1920s, *Women's Club* ran just one or two small sewing-school ads per month. By 1934, the magazine grouped anywhere from fifteen to twenty sewing-school ads each month into a several-page "School Guide for Women Looking for a Trade."[14] The magazines, for their part, ran articles with instructions on sewing the latest styles of women's or children's dress. *The Housewife's Companion* ran these pieces through 1944. As supplies of new fabric vanished, the magazine taught readers to recycle fabric from old clothes.[15]

The dream of some graduates reached beyond family sewing; these women hoped to go into business as home-based dressmakers or perhaps open a dress shop. A city-based dressmaker might take in two or three young girls as apprentices from the countryside, offering them room and board plus modest wages. The apprentice could hope to save a dowry while learning skills that would be useful both for earning money and managing a household. Even in the 1920s, a number of women had opened successful Western dress shops, but for the most part the public face of the garment industry was male. But through the 1930s and into the 1940s, women came to supplant men as the makers and to some extent the sellers of Western wear for women and children. In 1943, the Tokyo Association of Made-to-Order Children and Ladies Garment Businesses, as a condition of receiving rationed fabric, compiled what seems to be the first relatively comprehensive register of Western-style dress shops. It counted 1,282 retail stores in Tokyo alone. Women owned or operated 572 of them (45 percent).[16] At the height of the war, the value placed on economic independence for women remained strong, and growing numbers were able to pursue this goal.

With the growth of sewing schools and needle trades, and the turn by women to Western dress, the bottleneck that had restricted the sale of sewing machines began to ease, and rates of ownership spiked. The greatest increase came from 1935 to 1941, when Japanese households purchased nearly 600,000 domestically produced sewing machines in addition to about 114,000 Singer machines sold from 1935 until the company's virtual ouster from the Japanese market in 1938. The surge in domestic sales was led by Pine-Janome, Mitsubishi, and Brothers, but it was boosted by a host of smaller new entrants. By 1939, the leading trade publication listed thirty-three domestic makers of finished machines and fifty-seven parts manufacturers, and it reported 1,588 sewing-machine shops operating throughout the empire, including 163 in Korea and 82 in Taiwan.[17]

Adding these sales through 1941 to the stock of Singer machines sold through 1934 yields a total of 1.55 million household machines sold since the early twentieth century. Not all were still in use, but these were extraordinarily durable goods, and it is reasonable to estimate that by the eve of World War II, close to one in ten of the 14 million Japanese households counted in the 1940 census owned a sewing machine. Ownership by all accounts was significantly higher among city dwellers, and ownership overall had increased threefold in ten years. Several factors drove this spread: increased purchasing power as the economy recovered from the depression and real wages increased; lower prices offered by Japanese producers; and the desire of growing numbers of women to earn money as dressmakers or seamstresses, or to dress themselves and their children in economically homemade but fashionable Western dress.

Also important in the spread of the sewing machine and a growing part of economic life more generally in the 1930s was expanding consumer credit. In addition to items long available on installment plans—sewing machines, housewares, and men's Western clothing—goods newly available for purchase "on time" by the late 1920s included radios, pianos, automobiles, and suburban homes sold with multiyear mortgages.[18] In 1935, the City of Tokyo's Commercial Section undertook Japan's first systematic survey of installment credit. It found that ten thousand of the one hundred and thirty thousand retailers in Tokyo offered plans that financed 8 percent of the total value of the city's retail sales.[19] The top goods sold on time were men's garments, bicycles, automobiles, shoes, radios, sewing machines, books, medical and scientific equipment (including cameras), watches and jewelry, and Western furniture. The link between credit and a new consumer life tied to goods of Euro-American origin was clear; with the exception of books, every item on this list was a product of Western material and, for the most part, of industrial civilization.

Although a wide gap remained between modern consumer life as dream and as materially owned reality, it began to close during the 1930s as a growing proportion of people, especially in cities, came to possess the objects that defined middle-class modernity. An even greater number joined modern life simply by strolling city streets, reading monthly magazines, going to the movies, or listening to radio broadcasts ranging from lessons in English or sewing to Olympic sports. This expansion of material modernity took place at the very least *in spite* of the drift toward war, that is, despite the increasingly censorious demands of political leaders for people in a "time of emergency" to live more simply in the name of patriotic service and sacrifice. But this modern life also spread *because of* the mobilization for war. A radio might be purchased out of the desire to hear war-related news bulletins, but the radio also carried all manner of messages. The economy recovered rapidly from the depression, spurred in part by war-induced deficit spending, which gave more people the means to buy consumer goods and take part in new leisure activities. One attractive package that was part of this burgeoning modernity was a self-reinforcing set of desires and practices centered on sewing and dress: magazine spreads on the latest fashions, the growing popularity of sewing schools, the increased number of women dressmakers, the spread of homemade or custom-made Western dress among women, the expansion of consumer credit, and the growing sales of sewing machines.[20]

SMART AND SELF-RELIANT WOMEN
IN THE TIME OF EMERGENCY

As in the 1920s, the cultural appeal of this modern life derived both from the ebullience of film viewers and fashion seekers and a more sober ethos of the rationally future-oriented subject. One sees both aspects with particular clarity in Pine-Janome's brilliant advertising campaign of the 1930s. Its mastermind was Shimada Takuya, who began as the company's marketing consultant in 1934 and later signed on as a full-time employee. Shimada produced a remarkable set of text-intensive ads, horizontal boxes running the full width of the newspaper page, until they shrank in 1939 to small boxes with brief slogans. Shimada was also renowned for his commitment to scientific marketing; with statistics he sought to correlate sales results with particular ads. New Pine-Janome ads appeared every several days in the major newspapers from 1934 through 1941; the *Asahi* and *Yomiuri* papers each ran more than six hundred during this span.[21] They sometimes took the form of (purported) user testimo-

FIGURE 20. This Pine-Janome ad ran in the *Asahi shinbun* on January 9, 1936, just six weeks before the shocking coup attempt of February 26. As it exhorted women to save money with a scientific product and use those savings to shop without reserve, it brilliantly connected the scientific and rational aspect of modernity with the face of the pleasure-seeking consumer. (Janome Sewing Machine Corporation and *Asahi shinbun*)

nials; in other cases they offered advice to explain how the sewing machine could enrich a woman's life.

The sewing machine offered fertile ground for a campaign appealing to a protean modernity because it was both a tool of production and an object of consumer desire. It was particularly amenable to appeals to an economizing rationality that one might expect in an era of wartime stringency—for example a call to sew clothes at home rather than buy a ready-made dress or engage a seamstress. Certainly, Janome and other sewing machine sellers made such frugality-oriented claims for this good. More surprising is the presence, often in the same ads, of an exuberant, even defiant, celebration of the sewing machine as a means to live a modern, American-linked consumer life.

Consider, for example, this ad, which ran in the *Asahi* and *Yomiuri* in January 1936 (figure 20). Next to the jaunty profile of a Caucasian-appearing woman, the headlines read, "You! This year's resolution is to earn a ton with your woman's hand! It's shop 'til you drop, as fast as you can." The text goes on to explain that "the American woman is the one who plays more, buys more, and earns more than any in the world. In our country as well, in 1936, one new trend for women is to earn without hesitation and to quickly buy whatever you like without holding back." However, the reader is asked, how on earth can a woman with obligations to do much housework find time to earn and then to spend money? The answer is simple: "The problem is solved perfectly by bringing in science." If one buys a scientific product such as a sewing machine, one can economize on time and expenses and use the profit to shop to one's heart's content.[22]

Sewing machines were elsewhere aligned with a wartime rationale of

FIGURE 21. This Pine-Janome ad from the *Asahi shinbun*
of January 5, 1938, used a pun to connect the sewing machine
to the wartime causes of pronatalism and increased savings.
Its first two lines read *"Fume yo, fuyase yo"* (pedal and multiply), playing on the biblical phrase *"Ume yo, fuyase yo"* (give
birth and multiply) while echoing the state's wartime propaganda. (Janome Sewing Machine Corporation and *Asahi
shinbun*)

commitment to the state that left little time for a consumer life of pleasurable spending. Janome beat the patriotic war drum with ad slogans
ranging from the neatly gendered "For you a bayonet; for me a sewing
machine" to a marvelous play on the state's call for women to give birth
for the sake of the nation. Beside a biblically inspired picture of a snake
offering Eve an apple as Adam looks on, this ad exhorted women to
"pedal and multiply; save with a sewing machine" (figure 21).[23] This ad
acknowledged that people had to adapt to a new era of war, but even
here women were implored to build the nation and support the state
through the modern project of saving, as well as the ancient one of pro-
creating presented in Western religious phrasing and image.

　　Well into the war, Shimada's ads frequently positioned the sewing machine both as a means to save or to earn money in the causes of female
self-reliance and as a good bringing convenience and pleasure. In the summer of 1935, "the modern woman" was defined in a bold headline as
"one who can stand splendidly on her own, if need be." On February 24,
1936, Janome told readers to "earn with your hobby, Western sewing
and handicrafts." In an ad from late January 1938, a Western woman's
face is adorned with a college graduate's mortarboard, while the ad copy
offers a twin claim of security plus enjoyment amidst the uncertainty of
war: "With one [sewing machine], whatever happens, the life of a lady
will be pleasurable and free from care." In January 1939, sewing machines

FIGURE 22. Reminiscent of the lakeside sewing party of the Meiji court ladies of 1893 (figure 3), but considerably less of a fantasy, this photograph advertising a sewing school in the April 1933 issue of *Fujokai* fashionably conveys the pleasure of young-adult women stitching for themselves.

are still being described as "modern and smart." And on the very day that Japanese warplanes attacked Pearl Harbor, Janome unsurprisingly claimed that its customers could use sewing machines to save on clothing expenses. But Shimada encouraged women to use the money (and time) conserved not, as one might expect, to invest in war bonds, but to pursue the hobby of reading and the quintessentially individualistic goal of self-cultivation.[24] To the end of this prewar-to-wartime ad campaign, Shimada celebrated modern life and values.

Sewing schools echoed this marketing chorus with advertising in women's monthlies that similarly mixed appeals to pleasure, patriotism, and practicality, including the value of Western sewing skills in the marriage market. In April 1933, a photo spread in *Fujokai* presented an idyllic scene at the Mode Western Sewing Institute under the caption "These young *madames* and *mademoiselles* lose track of time in the pleasure of stitching these wonderful styles" (figure 22).[25] A twelve-panel comic strip the following January in *Women's Club*, resting atop the "School Guide for Women Looking for a Trade," offered one view of the future of these women (figure 23). It narrated the chance encounter of a well-off older

FIGURE 23. Enthralled by the prospect of a self-reliant bride for his son, the aspiring father-in-law in panel 10 of this January 1934 advertising supplement in *Fujin kurabu* dives into the doorway to a "school for working women," shouting, "If we take a daughter-in-law, she has to be a working woman!" The strip reads right to left.

couple, who had been chatting about their hope of finding a suitable mate for their son with a young woman on her way to a sewing class. The young woman's combination of manners and aspiration for personal advancement impressed the older couple so deeply that, in the final panels, the prospective father-in-law chased her into the school building. She won them over not with a demur expression of ladylike deference but a bold statement linking national crisis to women's individual aspirations: "This [learning practical skills like sewing] is the common sense of the woman of the new era.... A woman who cannot walk independently on her own cannot carry Japan forward in this time of emergency."[26] This fashion-conscious patriotism was echoed as late as December 1938 in an ad recruiting students to the "Ladies Western Sewing Academy" in *Women's Club,* which combined an offer of reduced tuition to women whose husbands or sons had gone to war with the proud claim that "our students get continuing reports of the news from Hollywood, the world vanguard [of fashion], from head teacher, Miss Opal Street.[27]

It is not easy to assess how young women themselves or their husbands and parents-in-law interpreted mixed messages about the patriotic pursuit of vocational skills, the pleasures of home sewing, and the excitement of shopping like an American woman. The graduation and classroom photos of sewing schools offer signs that the increased numbers of students at sewing schools in the 1930s included women who wanted to make West-

ern dress for themselves. Photos from the 1920s showed virtually all students in kimono, with only the teachers in Western dress; by the mid-1930s, roughly two-thirds of the students came to class or graduation wearing Western dress.[28] Signs of a desire for vocational training and economic independence are found in the growing numbers of women who engaged in dressmaking from the home or opened dress shops.

They are also found in the continuing and growing support for women's economic independence in the genre of "true stories" in women's magazines. One finds thirty-nine articles published in *The Housewife's Companion* between 1932 and 1945 that described the successful effort of a Western-dress seamstress, dressmaker, or dress shop owner.[29] These enterprising homeworkers included married women, single women not yet married, and war widows. As presented in *The Housewife's Companion,* they were often sewing to help their children move up in the world, indicating "supplemental income" as their main reason or writing explicitly that the primary motive was earning for the children's education. Growing numbers were prompted by a war-related circumstance, whether the draft or death of husband or parent.[30] More than half of the women worked entirely at home; the remainder were evenly divided between those who had opened dress shops and those who sewed outside the home for others (for example, in a small workshop in the neighborhood).

It is easy to imagine that far more numerous than those who opened their own stores were those who thought about doing this but stopped short and those who tried and failed. Less ambitious activity, such as home-based sewing for a broker or fashioning garments on order from neighbors, was featured in many tales, and indeed was more common.

A pioneering wartime study of time use in daily life carried out in 1941 and 1942 offers evidence that a wife or daughter undertook homework in roughly 3 percent of "salaryman" households and 13 percent of the homes of male factory laborers. About one-third of this work involved sewing or use of a sewing machine.[31] The survey recorded what appears to have been the peak of wartime homework. From the mid-1930s through the early years of the Pacific War, the pages of women's magazines presented a steadily increasing array of stories on how to find profitable sewing homework. In 1942 and 1943 every issue of *The Housewife's Companion* included pieces on both needlework and other homework opportunities. Such articles and, one assumes, the ability of contractors to offer these jobs diminished in 1944 and vanished in 1945.[32]

Other sources offer a sober view of the prospects for these women. In February 1935 *Fujokai* began a long story on homework with this caution: "The difficulties of life are only increasing. Wives cannot simply sit and wait for husbands to bring home the family income; . . . more and more women are looking for homework to help the family budget. Putting aside shame, struggling seriously for a livelihood in this way is a good thing." But, "as those in need increase, so do the cheats lying in wait to take advantage of them. One finds so many cases these days where a woman lets down her guard, whether from greed or anxiety, and is caught helpless by a fraud."[33]

The safe and recommended course of action was to seek vocational training, most commonly machine sewing, through training centers set up by municipal governments or Patriotic Women's Federations. These centers (*jusanjo*) had their origins in municipal social welfare programs initiated in the early 1920s that had expanded significantly from the depression through the mid-1930s. Eligibility was partially tied to need, although women of some means were able to find work through the centers. Training was practical and limited; a woman was taught only what she needed to know for a particular job—for instance, how to stitch together certain cuts of fabric. One could expect to earn from five to thirty yen a month. The report concluded cautiously that a woman could not really support herself or her family on such an income but could make a good supplemental contribution to the household.[34] Another report from the daily press noted that about half of the jobs involved machine sewing. The women who took them were not the poor, but the wives in "homes of the broadest social strata of all, the lower middle class." They typically added a significant 30–50 percent to household income.[35]

Such data along with the large number of female-owned dress shops registered in Tokyo in 1943 make it clear that *The Housewife's Companion* was not spinning pure fantasy with its "true tales." Certainly the magazine idealized the independent female entrepreneur, but we can see this ideal as a powerful cultural reality sitting uneasily alongside a messier social reality.[36] *The Housewife's Companion* sought to inspire readers with true tales of women who achieved economic independence, security, and a measure of happiness—in some cases, to be sure, after a few years of struggle. The proportion of shop owners in these tales was higher than in the general population of dressmakers. In nearly half of the magazine's cases, the dressmaker or seamstress accounted for more than half of her household's income. In the magazine's telling, homework or a dress shop offered more than a supplemental income to a struggling family; it was a means to female independence where the wife earned more than the husband, and a family might secure a comfortable, middle-class life.[37]

Homma Yuriko of Osaka, for example, told of "sewing homework that earned children's school costs." Her efforts began in 1933 when she started taking in sewing from a local fabric broker. She was soon earning more than forty yen a month and exclaimed that "there was no more problem in paying my son's and daughter's school costs. Is there any greater joy for a parent, for a mother? My husband [whose own business venture had failed] was also so happy. The once dark atmosphere at home felt suddenly brightened. . . . The sewing machine which I had earlier bought simply for family use, when my daughter entered school, has become an important life resource that rescued us in a time of trouble; the Western sewing that I learned for the fun of it has been so useful." From homework, she and her older daughter turned to making children's and women's clothes on order for neighbors, "with reference to foreign catalogs and the patterns published in the supplements to *The Housewife's Companion.*" When her younger daughter sported her homemade "new style sailor suit" at a school field day, other parents came to order "that same cute outfit."[38] Here was the ideal of modernity in the "time of emergency": a celebration of investment in the sewing machine as a capital good, the careful management of the home, the rational deployment of resources, but also the pleasure of keeping up with fashions and changing lifestyles.

REGULATING MODERN LIFE IN WARTIME

In ads for sewing machines and schools, the juxtaposition of calls for self-cultivation or feminine pursuit of pleasure with exhortations to save and serve the nation at war continued until the eve of the Pacific War. The attraction of Hollywood as the vanguard of modern culture persisted as well, even after the Ministry of Finance began to restrict imports of foreign films in 1937, and the Film Law of 1939 escalated state censorship of films and other media to insure that their messages were properly patriotic.[39] The number of moviegoers continued to rise, and as late as December 4, 1941, officials cleared ninety-six American films for immediate release.[40] Even as battles raged in China, plans were finalized for the attacks on Malay and Hawaii, authorities were encouraging sacrifice for the state with slogans such as "destroy the private, serve the public," and in the face of growing scarcity, consumer desires persisted. With a notable dampening, to be sure, this spirit continued well into the Pacific War. In many realms, including that of sewing and dress, it was not until 1944 and 1945 that a nightmare of deprivation and then physical destruction squashed virtually all leisure and discretionary consumption.

In this embattled context, modernity was confronted not so much by premodern tradition itself—defined as long-standing customary practices in daily life—as by a modern spirit of "traditionalism." Agents of the state as well as some of their foes on the civilian right loudly propagated the belief that time-honored, quintessentially Japanese practices, which were in large measure invented or cobbled together in the modern era, should be touchstones for morality and action. The young military officers who assassinated Prime Minister Inukai in May 1932 had originally planned to kill both Inukai and Charlie Chaplin at a reception honoring the latter on his visit to Japan, until last-minute scheduling changes led the rebels instead to attack the prime minister at his residence.[41] The assassins condemned modern Western culture as decadent and sought to restore imperial rule, but they stood on modern ground as they advocated universal suffrage and the nationalization of big business. In the following years, a narrowing traditionalism promulgated in key texts such as the 1937 "Fundamentals of the National Polity" echoed the rhetoric of the rebels. This 1937 manifesto from the Ministry of Education exalted the emperor and blamed Japan's crisis on Western beliefs ranging from individualism to communism. In their place, it exalted loyalty and military spirit as the nation's core values and a hierarchical family system as its core institution. Even so, its underlying logic rested in Western thought, specifically a Hegelian idealization of the nation and state.[42]

Like such explicitly political traditionalism, the more implicit politics of daily life and popular culture involved a tense dance between those who advocated frugality, restraint, and the purging of Western modernity from a purified Japanese national body, and those who—though patriotic supporters of empire and emperor—were unwilling to jettison the habits modern life. In August of 1940, Tokyo city officials placed fifteen hundred signboards painted with the slogan "Luxury Is the Enemy" at strategic locations throughout the capital. Dance halls were banned that year as well. After the Pacific War began, a war on baseball as an enemy sport and jazz as enemy music opened on the home front. But where the means to consume or to play could be found, slogans and regulations did not effectively change behavior. In 1939 the government advised local authorities to exercise "guidance" to curb the fashion for permanent waves and other corrupting extravagances, but beauty parlors continued to draw customers. Not until the second half of 1943, and only after prolonged negotiations with beauty parlor owners over the price they would receive for hair dryers whose metal parts would be recycled for military use, did authorities ban the use of electric hair dryers. The decision was aimed less at dictating style than at conserving electricity.[43]

The purchase of goods with money yet to be earned, in the form of installment payments in particular, likewise continued in the face of an anxious mix of criticism and support. The view put forth in the late 1920s of consumer credit as teacher of the rational, modern discipline of budgeted expenditures and provider of the fruits of industrial civilization persisted through the following decade. When the city of Tokyo published its study of retail installment selling in 1935, the Tokyo Asahi gave it positive play with a headline proclaiming "Living Standards Rise [even though] Income Stays Put." Ogawa Nobuo, the city official who conducted the survey, told the reporter that with ten thousand retail sellers offering their goods on time, today's middle class "wears installment from head to toe" and then some; they bought furniture and homes on credit as well. In his view, despite lingering scorn in some quarters, the image of installment goods as shoddy and overpriced had largely been overcome.[44]

But an undercurrent of concern with the perils of borrowing was present as well. In the daily press it appears that upbeat views such as Ogawa's were less common than anxious ones. In fifty-five articles on "monthly installments" in the Asahi from 1926 through 1941, a strong minority of nineteen stories presented credit purchasing as a promising or at worst neutral practice. A slightly heavier dose of twenty-three stories offered sensational reports on various credit scams and scandals that rendered installment selling as problematic.[45] A 1933 piece, noting that among the "salary-man installment class" everything from clothing to furniture to homes was bought on time, cautioned in its headline, "No Room for Carelessness." Later that year, readers were told that while fraud was less common these days, buyers needed to remain vigilant to avoid overpriced deals or shoddy goods.[46]

The relative balance of positive and negative *Asahi* reporting on consumer credit did not change dramatically over the fifteen years from 1926 to 1941, and the balance of coverage in seventy-one installment-connected stories in the *Yomiuri* over this span was similar.[47] But toward the end of the 1930s, some of the negative stories took on a newly censorious, preaching tone. The final article in a four-part series in the *Asahi* in 1938 on trends in consumer finance, although headlined "Monthly Installment Payments Convenient for Salaried Workers," concluded that "it is not pleasant to borrow money, and we are better off living within the limits of our incomes and saving to meet emergencies, building a truly strong livelihood plan that way."[48] And in *Advertising That Works,* a book published the same year, a well-known expert on marketing, Matsumiya Saburō, offered a far less measured blast: "The American-style system of installment selling is extremely cruel and completely at odds with the spirit of the Japanese nation and the virtue of Japanese people." He told a sad story of a friend victimized by Singer Sewing Machine Company's punitive installment contracts. He concluded that "our country's origins lie in the imperial spirit, and we have developed as the land of the gods. America is an immigrant nation, a collection of people from all over the world, where nothing is as respected as money. It is a country that worships only lucre as its true faith. Our country has a completely different spirit."[49]

A combination of such criticism and growing state regulation of the economy eventually forced sellers to curtail consumer credit. Even so, installment plans remained attractive and were even appropriated at times for the war effort. In August 1939, the *Yomiuri* proudly launched a Yomiuri Welfare Fund, providing interest-free, unsecured loans to needy families of active-duty soldiers for unusual expenses such as births, weddings, or funerals. The paper boasted, as did the ads for radios or sewing machines, that "this is easy money," available with minimal red tape, repayable in monthly installments.[50] A few months earlier, the *Tokyo Asahi* reported that in response to police "requests" the two thousand members of the Tokyo Association of Tailors were taking two steps: lowering prices of suits by 5 percent and ceasing to offer installment credit. The headline stressed not the benefit of the lower price, part of the government's effort to control inflation, but the "Greater Pain for Salaried

People" caused by the end of installment sales.[51] The pain was apparently sufficient to lead the authorities to change their position. In July of 1941 the Ministry of Commerce authorized the sale of custom-made Western dress on the monthly installment plan, with a ceiling of a 5 percent interest charge spread over six payments.[52]

Amid this swirl of mixed messages, with luxury the enemy and Hollywood the dream, with credit a means to survive or live a good life but a peril of tainted origin, consumer life persisted and even thrived through the end of the 1930s. But with 1939 as one turning point and 1942–43 as a more decisive break, increasingly stringent economic controls along with a growing scarcity of resources severely limited the ability of women and men to pursue their modern pleasures. To hold down inflation, the government imposed price and wage controls in 1939 and established rules limiting the free movement of workers. These policies were designed to end the "wasteful" competition of a market economy and to funnel capital and raw materials to military production. They made it increasingly difficult to manufacture and sell consumer goods, and the sewing machine was no exception. The price-control order of September 1939 froze sewing machine prices at a time of rising material costs.[53] The order initially allowed goods to be sold on installment for a maximum of 15 percent above the cash price, and Janome continued to offer household machines on the installment plan. But the government would not allow the company to continue its popular layaway plan.[54]

Despite such restrictions, domestic makers managed to increase production and sales of home-use machines for a time. Sales reached an all-time peak at 154,000 units in 1940, with a slight decline to 142,000 units sold in 1941. The number of producers continued to grow as well. Only with the start of the Pacific War did more draconian policies directing all industrial production to war goals dramatically constrain production for consumers. In February 1942 the Ministry of Commerce began a push to consolidate the machine industry, including sewing machine producers. In their case, only plants with a minimum of one hundred production machines would be allowed to remain in operation; parts makers had to maintain factories of fifty or more machines. These limits were harsh; at the time of the consolidation order, only nine producers of finished machines and twelve parts makers met these standards. Hundreds of smaller firms were required to amalgamate to reach this scale or leave the industry. Many took the latter course; production of household machines declined to 51,000 in 1942 and just 16,000 by 1944.

The regulations at first glance worked as intended, in that many producers shifted from making sewing machines or parts to become contractors to large munitions producers. But a naked army could not fight. Increased numbers of industrial-use machines were needed to make coats, gloves, uniforms, boots, or parachutes for the soldiers. Many of the machines sold in the "household" category were surely used for this sort of production by homeworkers or in small garment shops, but it is striking that the state did not prioritize production of industrial-use sewing machines to supplement or supplant home-based contracting. The production of such machines was low to begin with and rose only slightly from 2,900 in 1941 to 5,400 by 1944. This industry—like all others—was literally demolished by firebombing in 1945, when only 4,000 household or industrial machines were produced.[55]

Janome continued to sell some of its machines on the installment plan until March 1942, when it discontinued the practice—apparently the last seller to do so. Rather than a government order, it was the combination of scarcity of materials, falling production, shrinking profit margins, and rising prices that generated too much uncertainty to allow the company to offer credit.[56] No installment-plan ads for sewing machines appeared in newspapers after 1942. A handful of ads for cameras and radios sold on time ran through the end of 1943, when they vanished as well.

FABRICATING A WARTIME JAPANESE MODERNITY

To tell the story of wartime modernity in daily life with a linear narrative of the surprisingly late and slow but nonetheless inexorable constriction of consumption and leisure would be a revisionist project of some value. But to end matters there is ultimately misleading. Such a framing fails to capture the complexity of negotiations to control tastes across realms from music to hair styles and dress that continued even through 1944, and it misunderstands their outcome. In *Blue Nippon*, Taylor Atkins argues that the persistence of jazz in wartime Japan sparked creative efforts, however riddled with contradiction, to devise an "authentic" Japanese jazz.[57] Echoing these improvisations, those who designed clothes, taught sewing, or used sewing machines took part in a tortured wartime effort to design an "authentically" Japanese but functionally modern dress. These efforts to define a Japanese life in a time of total war, one can say—at some risk of slighting its ultimate horror—were ironically productive. Amid a flood of rhetoric that condemned the modern and the West for corrupting the purity of the Japanese nation and soul, people searched for an "appropriately Japanese" modernity. The prolific ethnographer of modern life, Kon Wajirō, and the energetic sewing ped-agogue, Narita Jun, both joined this effort. Their shifting and different perspectives offer insight into the complicated alliance that drove the war-time efforts at dress reform.

In 1937, just months before the outbreak of full-scale war in China, Kon began to shift his fashion enthusiasm in a short review of Japanese and Western dress. In the past two or three years, he observed, the fresh-ness of Western dress fashion had declined, and its form had become "relatively static," imitative, and derivative. Fashions in Japanese dress, in contrast, had been creative and innovative, thanks in part to the "un-yielding efforts of our country's weavers." He concluded with a ringing flourish that with Japanese dress "liberated from the restrictive conven-tions imposed since late Edo times," the present was a time of dynamic advance fully deserving the label of "renaissance."[58] Kon's language was artful in using the Western-derived word *renesansu* to draw Japanese dress into the charmed circle of modern trends.

Nine months later, in a piece for the Osaka Mitsukoshi Department store magazine that would have been written during the massacre in Nan-jing, Kon had sharpened his analysis. He produced a nuanced but (to present-day readers) troubling assessment of the cultural dynamics of modernity and tradition, East and West, and imperial power. Picking up where he left off the previous May, he noted that contrary to conven-tional wisdom and despite the purported "rationality" of Western dress, younger Japanese women had recently come to a new appreciation of the beauty of Japanese dress. Although pleased at this trend, he took is-sue with the approving remark of a French "culture critic" who, on a re-cent visit, had pronounced Japanese women fortunate to have the choice of two dress modes. In a prescient critique of what has come be called orientalism, he wrote that it was unsurprising but still "insupportable" that "foreigners as they rush to make cross-national comparisons, are generally satisfied to take pleasure in the sight of a fixed, stagnant, Japan. They are not happy to see fashions change, but look for fixed ethnic pat-terns."[59] Kon admitted that for quite some time, Japanese dress had been relatively fixed. But as in Europe after World War I, a newly innovative sense of fashion had emerged in recent years. Kon welcomed this spirit, which he understood to have been stimulated by the desire to devise a Japanese form of dress that made sense in the context of the modernized (that is, Western) dress, architecture, and transportation in Japan.

For Kon, "Japan" and the "West" were complex formations that de-fined each other as they both changed. One is impressed, to this point, by his ironic and critical stance, his awareness of the power relations embedded in cultural interactions, and his resistance to a reductive understanding of fixed Japanese and Western essences. This makes his concluding comment all the more demoralizing. "The mood of the so-called time of emergency," Kon wrote,

> brought with it a sort of oppressive darkness, but recently even this state of emergency has, by the might of the emperor's military, been transformed and bathed in a positive spirit. The emergency has brightened; there is room to relax and to persevere in expectation of success. This changed mood among people, not the result of anyone's planning, manifests itself in expressions of new fads and fashions of the season. The weavers, the sellers, the buyers, the wearers all join in this mood and new fashions are born.
>
> Surely, one imagines that in upcoming fashions we will find brightness and themes that powerfully exalt and sing the praise of our nation. And rather than looking back toward a national style, these fashions are likely to be of a character that contains elements of strong international appeal.[60]

Kon was just one of many sophisticated observers who lost his bearings in a time of war. Despite conducting ethnographic surveys in the colonies as well in the home islands, Kon utterly failed to recognize his own implication in projects of aggrandizement that imposed the will of the Japanese state on others. Yet he was far from a narrow-minded or naive defender of a timelessly authentic Japanese spirit. He skewered the imperialist cultural stance of others, and he celebrated an expected Japanese contribution to a global modern culture. He was an accomplished intellectual voice of wartime modernity.

Just as Kon in the 1920s had begun his observations with excitement over the spread of Western fashions, Narita Jun had been among the many sewing pedagogues who embraced Western dress and sewing, although she did so to defend sewing as a home-based and essentially feminine pursuit. In the changing political and cultural environment of the 1930s, she and her colleagues in sewing pedagogy continued their impassioned debates. Some argued that women should be taught sewing as a vocation that might support a family. Others, Narita prominent among them, saw their mission as the teaching of "housewife sewing." But even as they rejected the teaching of sewing as a vocational or commercial pursuit, such educators were sharply divided among themselves. Those we might see as more reform-minded and modern, such as Ushigome Chie, introduced above, or Sakai Nobuko of the Nara Normal School for Women, saw that middle- or upper-class housewives were increasingly buying ready-made wear or ordering it from dressmakers, and they argued that for such women, classes on sewing were too narrow. Sakai in her 1937 book, *Basic Principles of Sewing Pedagogy,* called for holistic training in "clothing" education. "Even if they do not make clothes themselves, [students] will need to know how to buy the materials [for others to make], how to select fabrics, how to purchase ready-made dress. . . . This sort of knowledge will be important no matter how widespread ready-made dress becomes, and will defend the sewing curriculum from a charge of 'irrelevance.'"[61]

Narita was a leader among those taking a third direction, neither vocational nor holistic. Around 1933 she moved away from her earlier promotion of Western dress and sewing as the way to nurture home-bound feminine virtue. She began to criticize modernist fashion and the adoption of Western dress by urban women as a superficial and frivolous imitation of the foreign. She called for education emphasizing "Japan's uniqueness."[62] In a 1936 essay entitled "How We Should Think of the Future of Sewing [Education]," Narita gave pride of place to hand sewing and Japanese dress in fundamentally moral rather than practical terms: "When we give heart and soul to sewing stitch by stitch, we feel an inexpressible joy. Femininity and gentleness are somehow nurtured, giving birth to a womanly consideration for others. . . . After all, the right role for a Japanese woman is to protect the home."[63]

One might expect state officials, as they sought to dress the nation's women and men for war in a fashion both efficient and economical, to have enlisted as their primary allies those in the holistic or vocational camps, people more practically oriented than Narita. To be sure, the Ministry of Education in 1943 did broaden the "sewing track" to a "clothing track" in the first significant revision of the home economics curriculum for female students in decades. To this extent, the authorities sided with Narita's opponents, such as Sakai, by endorsing education for women not simply as fabricators of dress but as more comprehensive wardrobe managers.[64] But the misfortunes of war offered Narita an opening to make her case. A growing scarcity of both fabric and new sewing machines meant that her stress on home sewing and hand sewing, while driven by a moralizing impulse centered on feminine values and aesthetics, converged with the state's stress on the rationality of efficiency and convenience. Rationing of fabric for civilian use boosted the attraction to authorities of Narita's "home sewing" ideology, which promoted recycling and repair of old clothing. She took a prominent part in the state-led effort at wartime dress reform, serving on the committee that selected designs for women's "Standard Dress" in 1942.[65]

Narita was not part of the wartime project of clothing reform from the outset, for just as the Meiji government in the nineteenth century promoted men's Western dress well before that of women, the state's wartime drive to reform the dress of its subjects began with men. In late 1937 the Cabinet Information Board announced a plan to devise what it called *kokumin fuku*. The term translates as "National Dress" with no gender marking, but this was in fact a men's outfit; the label carried the none-too-subtle meaning that men constituted the "nation."[66] It took a year for the Ministry of Welfare, in November 1938, to start the project in earnest by convening a "dress committee" with eighty-three members from the Ministries of the Army, Navy, Commerce, Agriculture, Education, and Transportation, as well as journalists, youth-group leaders, educators, and other experts.[67]

Although he did not join this committee or take other official roles in the state-led projects of wartime dress reform, Kon Wajirō offered his qualified endorsement of dress reform in two essays at the start of 1939. He noted that economic pressures were forcing many families to economize on clothing, repairing, and recycling the wear of adults in particular. He recognized a need for the garment industry to design economical and practical clothing, but he opposed a one-pattern-fits-all national uniform as stultifying progress in dress. He called for different uniforms for different social groups.[68]

The path toward an authorized design was winding, but it ultimately betrayed Kon's hope for uniform variety. The project was briefly suspended in 1939, and later that year the army took charge, convening a much smaller Dress Association, just sixteen in all: a few civilian experts and representatives from key ministries. The committee invited the public to submit designs in a contest to select this new national menswear. The varied criteria were impossible to satisfy fully. National Dress was supposed to be brown, appropriate for daily wear but easily convertible to a military uniform; not dramatically different from existing [Western] dress (to allow men to convert their current clothes to the new design); healthy, economical, and appropriate for physical labor. It was also expected to "bring to life the special character of Japan's particular dress, [and] be distinctive and progressive" while serving as "a Japanese outfit that can play a leading role in world clothing culture."[69] Toward the year's end, two national daily papers cosponsored the month-long contest, which drew only 282 entries. The committee modified some of its favored submissions, and in January 1940, with much fanfare, announced four authorized outfits.[70]

These garments were poorly received, despite government promotions featuring cabinet ministers wearing National Dress to work and pressure on businesses to require employees to wear the outfits or, better yet, purchase them for their workers, and despite claims that they saved material and cost half the price of a typical suit. In November 1940 the government consolidated the four designs into two quite similar National Dress types, A and B. It issued a National Dress Order that declared these outfits "examples" of what men ought to wear but did not command men to adopt them.[71] In November 1943 the two types were consolidated to one design, but despite continuing efforts at persuasion, a 1943 survey by Kon Wajirō's long-time collaborator, Yoshida Kenkichi, showed that no more than 12 percent of men wore National Dress, while 83 percent still went about their business in Western dress. A mere 5 percent of men wore Japanese-style dress. Only in 1944 and 1945, as daily life became a quasi-military cycle of work and air-raid drills, did a large majority of the male population don these outfits.[72]

Not until 1941, more than a year after the rollout of National Dress for men, did the effort begin to design what came to be called "Standard Dress" for women (hyōjun fuku). While National Dress was promulgated with a state order carrying the force of law under the National General Mobilization Act, Standard Dress was announced simply by a vice-ministerial "Memorandum of Understanding."[73] But even though the state was late to the project, with this lower profile of action, the question of women's wear generated a more intensive and varied debate.

From the outside, Kon Wajirō endorsed the project early on. Already in 1939, as the drive began for men's National Dress, he had backed away sharply from his recent enthusiasm for fashionable innovations in Japanese dress, lamenting the hopeless rigidity of wafuku, which he blamed on the conservatism of sewing education in girls' schools. The only way forward, he thought, was for women to take Western dress as a baseline. From it they should adapt "a new people's dress for our nation."[74] By 1941 he had shifted his position once more. Present-day Japanese dress had some beautiful elements, but was not practical at a time when saving and conservation were top priorities. But Western dress was equally problematic. In Japan it was too much the domain of elites, because few in the "middle class and below" owned sewing machines or possessed the skills to make Western outfits. Wartime dress reform should not only aim at economic and healthful wear, but should be directed from "on high," integrate social classes, and bring the nation together.[75]

Although Kon did not take part, the Ministry of Welfare reflected the spirit of his suggestion when it began the Standard Dress project by convening an advisory board of experts in March 1941. Narita Jun was included among its members. The initiative shared several goals with the drive to clothe men in National Dress: conservation of resources (fabric), promotion of healthy bodies, and ease of movement for wartime labor. Unlike menswear, the women's outfits were neither limited to brown nor required to convert easily to military uniforms; in addition, they were expected to meet two criteria not put forward for menswear: to be "compatible with dwellings" and "easily fabricated in the home."[76] This final goal reflected a clearly gendered difference in the clothing supply chain. Women's wear—whether Western or Japanese style—was most often home-sewn; menswear was usually tailor-made or purchased ready-made. It also meshed with the belief of Narita Jun and others in the moral value of home sewing.

A contest was announced in October 1941 to select the best Standard Dress design. It drew 648 entries, far more than the menswear competition. The committee announced winners, including designs submitted by individual sewing school students as well as by sewing schools, which used the recognition to attract new students. The committee turned these designs over to its expert members for final modifications. In February 1942, to only modest media fanfare, several authorized styles were made public.[77]

Believing, it seems certain, that women were more concerned than men with choices in dress, the state offered no less than eight variations of Standard Dress, divided into two basic categories of Western- and Japanese-style standard wear, plus a third category labeled "active" dress. This last was in essence the work trousers called *monpe* (figures 24 and 25). The Western and Japanese modes were divided into A and B types, and for the Western mode, variations numbered 1 and 2 were offered for both the A and B types. In addition to these six options, the option of "active dress" *monpe* work pants was offered in two styles, a baggier Japanese mode and a trimmer Western type.[78]

Over the following months, the reformers who had come together in this military-bureaucratic-civilian alliance promoted Standard Dress through lectures, newspaper and magazine articles, and pamphlets distributed nationwide. To advocate the new mode, they organized a Greater Japan Women's Dress Association under the umbrella of the Imperial Rule Assistance Association. But despite these efforts, a column in the journal *Ifuku kenkyū* (Clothing research) had to admit in a 1943 article that "the numbers of women wearing pants in public has greatly increased. . . .

FIGURE 24. This photograph ran in the *Mainichi shinbun* on December 20, 1941, offering readers a preview of the tentatively approved designs for Standard Dress for women, officially promulgated in February 1942. Similar photos appeared in all the major daily papers. The photo shows a *wafuku* option on the left, two of the Western-dress choices in the center, and on the right the "active wear" *monpe*, which, contrary to the intent of most Standard Dress advocates, emerged as the most common wartime women's wear. (Mainichi Photobank).

One cannot deny that this is a sort of wartime fashion. We feel there is a very intriguing problem in the fact that compared to Standard Dress, which does not spread without vigorous promotion, women themselves are turning to pants, which have been encouraged by no one."[79] Dress historian Nakayama Chiyo echoes this contemporary view: "Almost no Japanese women adopted either form of Standard Dress."[80]

At first glance, it is puzzling to read this 1943 lament or understand Nakayama's similar point. After all, "active wear" *monpe* trousers were one of the three categories put forward in the announcement of Standard Dress. What these statements reveal is that state officials and their civilian allies who joined in the dress campaign saw their primary goal

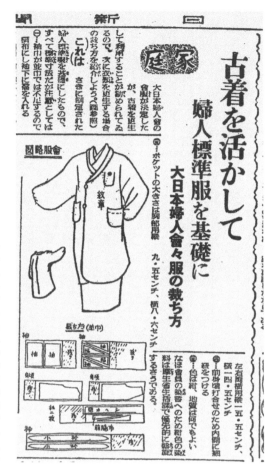

FIGURE 25. On May 20, 1942, the *Asahi shinbun* ran this article offering detailed instructions on how to sew a modified standard dress outfit, which the Greater Japan Women's Association (Dai nihon fujinkai) had adopted as its uniform. Women were encouraged to recycle fabric from other outfits to make this garment. One can also read the announcement as a modest wartime attempt to allow women some room for improvisation in their dress. (*Asahi shinbun*)

as promotion of the two categories of Western and Japanese style Standard Dress. They had added the category of "active dress" slacks with considerable unease to co-opt a trend already underway, a wartime search for both function *and* fashion. In subsequent discussions of women's dress, *monpe* were typically understood as separate from Standard Dress, even as a sign of resistance to it; a 1943 *Yomiuri* editorial, "The Question of Wartime Dress," condemned women for irresponsible use of expensive, gaudy fabric, "whether for Standard Dress or for *monpe*."[81]

Monpe was an umbrella term that referred to a wide range of agricultural work pants. Some in the Daily Life Reform movement of the 1920s had supported them as an alternative to Western dress, suitably functional for the domestic labor of modern life but suitably Japanese in origin. In 1930, an army paymaster, Kimura Matsukichi, contributed a prescient article on *monpe* to one of the first issues of the journal *Hifuku* (Clothing), published by the reform-oriented Japan Dress Association. Kimura located the origins of *monpe* with misleading precision in the Yonezawa farming region of northern Japan. Reflecting a military concern for mobilizing women as factory operatives in a possible future war, he concluded that *monpe* had much potential as work dress for women. They were "very easily sewn," inexpensive, mobile, and (for unexplained reasons) they "protected women's chastity."[82] Later in the 1930s, Segawa Kiyoko, a student of the ethnographer Yanagita Kunio, traveled the Japanese countryside to catalog what she believed were vanishing dress customs of continuing value. She recorded more than three hundred variations of *monpe*.[83] Through such efforts, modern men and women were inventing the concept of "traditional" farm dress as they sought to modify, appropriate, or defend these work pants.

Support for *monpe* as appropriate wear in a time of emergency intensified with the outbreak of the China War in 1937, nearly four years before the state began its formal project to design Standard Dress for women. Women's magazines began to solicit design ideas for "emergency era clothing" from readers.[84] In 1938 the venerable Patriotic Women's Association (Aikoku fujin kai), linked to the Home Ministry, convened a "*monpe* workshop" at a girls' school in Tokyo that taught women how to make and wear this dress and extolled its value for firefighting.[85] In 1940 the advice column of the *Yomiuri* "Women's Page" sought to answer those who complained that, contrary to the claims of *monpe* advocates, these pants were in fact cold in winter and clumsy for work. The problem, the column countered, was that women did not wear them properly.[86]

As this defensive stance suggests, *monpe* had its share of detractors, including some in high places. The men who oversaw the army-connected National Defense Women's Association (Kokubō fujin kai) were distressed in late 1937 to find women in Osaka attending firefighting drills wearing *monpe*, and they advised against the practice.[87] The most important advocate of dress reform in the Ministry of Welfare, Saito Keizō, declared in 1939 that "recycled national defense dress such as *monpe* is a disgrace to the nation." In the discussions of the Standard Dress advisory committee in March 1941, a number of participants described their effort as aimed to halt the spontaneous spread of *monpe*. The ensuing vigorous debate made it clear that the slacks were already spreading from farms to women in cities and towns. For some of the men, the sight of women in pants was troubling because it erased their feminine appearance. For others, *monpe* raised alarm because they were allegedly unsafe; the long pant leg could too easily catch fire.[88]

The popular response to *monpe,* like the official one, was by no means unanimous or easy acceptance. On the eve of the drive for standard wear, these pants had been under discussion in clothing circles for some time, and some city women on their own had started to wear them, but no consensus had emerged as to their place in daily life outside the farm. In some quarters, a reputation not only for ugliness but clumsiness persisted, another sign of the ongoing wartime concern for fashion as well as function. The well-known blues singer Awaya Noriko recalled in her memoir that "even at the peak of the war, the modern girl would never ever wear *monpe.*" Yet for other women, writing at the time or retrospectively, these pants were a convenient, warm, and rational choice.[89]

One finds, then, through the end of the war, a swirl of opinions on how women should dress for service on the home front. Many of these came together in typically imperfect harmony in a 1944 book by Iwamoto Motoko titled *The Essence of Clothing: Women's Standard Dress.* Iwamoto was a professor of home economics at her alma mater, Japan Women's College, and a member of the Ministry of Welfare's Dress Advisory Committee. She called for a "self-consciously Japanese dress, a step advanced beyond Western dress." Such clothing would take in the "scientific" and "practical" aspects of Western dress, yet would respect traditional form expressing Japanese-ness, especially in the clothing of the upper body. One can read Iwamoto's statement as typical of the ultimately failed advocacy of a "Japanese modern" mode of dress, which, as Inoue Masato persuasively argues, carried forward from the Daily Life Reform efforts of the 1920s into wartime.[90]

Contra Iwamoto, and to the dismay of supporters of Standard Dress, *monpe* slacks, whether sewn with a Western cut and matched with shirts or fashioned in looser Japanese style with kimono-type upper garments, won the day as Japan's wartime garb for women. *Monpe* fit the bill as a properly indigenous mode for mobile activity. They fit Narita Jun's vision, for they could be sewn at home, whether by hand or by machine. Postwar citizens carried forward bitter memories of the deprivation and losses suffered while wearing them, but at the time of their adoption, reactions were far from uniform. They spread neither solely nor primarily by command, but rather by word of mouth and by example among women who were collectively improvising responses to their desperate situation amid a mix of messages.[91]

· · ·

In the process of fabricating wartime modernity, the ongoing ideological pull of rationality, centered on economy and efficiency in motion as well as in cost, is clear. Other combatant states, such as Britain, with its "Make Do and Mend" campaign, similarly promoted the efficient use and reuse of fabric and clothing.[92] But while issues of cultural or national self-definition were ever-present as states and societies sought to legitimize and mobilize for war, people in Japan faced with a particular intensity the question of defining and defending a cultural difference while embracing the global modern. Even as the West remained a persistently attractive source of "smart" modern styles, so did the desire to define an alternative Japanese version of the modern remain irresistible. Such desire was often combined with hope that Japan's version would both resist an arrogant West and contribute something distinctive to world culture.[93] The efforts of the state to pursue an indigenized modernity were extensive, but the state's reach had limits. Wartime modernity emanated from multiple places, including public intellectuals, commercial media, commercial schools, and women's organizations.

In the story of wartime sewing and dress one sees neither a straight line of continued Westernization of dress nor a clear move back toward "traditional" garb. Changes proceeded in fits and starts and with unintended consequences. In global comparative perspective, it was extremely unusual for so many women to so quickly put on slacks.[94] It is no surprise that the move to *monpe* provoked a variety of responses, including concern among women as well as men over compromised femininity and style. This change in dress was radical: a form of women's dress that until the late 1930s had been limited to farm work was adopted within a few years by virtually all women in their daily lives. It was also destabilizing. Although *monpe* as a sort of fashion choice prevailed over the even less popular Standard Dress, women abandoned them with little regret soon after the war. Their short-lived ascendance over "reform dress" alternatives discredited decisively the effort to design a functionally modern hybrid based on the kimono. This in turn cleared the way for a wholesale turn to Western modes after the war.

But one important practice *was* sustained and, if anything, intensified during and across the war: home sewing for long hours, whether by hand or machine. In 1941 and 1942, the national broadcasting network, NHK, conducted a massive study of time use to learn how it might align its broadcasts with the daily routines of listeners. As one unintended consequence, the questionnaires completed by nearly forty thousand individuals, divided into categories of farmer, factory worker, salaried worker, and small business owner, have left to historians a remarkably detailed picture of daily life in wartime. Women in nine thousand farming households reported sewing an average of 54 minutes a day. Women in working-class families averaged between 3 and 3.5 hours of sewing per day, while those in "salaryman" homes spent a little more than 3 hours at needlework.[95] These totals *excluded* home-based commercial sewing.

Like surveys from the early postwar years noted in the introduction, NHK's wartime inquiry did not distinguish hand from machine sewing. Nor did it separate Western from Japanese modes of stitching.[96] These extraordinary totals reflected all types of sewing done by women for themselves or their families. It appears that among this mass of household sewers, the minority fortunate to own a machine used it happily or proudly, while neighbors looked on with envy or desire as they stitched by hand. In more than one memoir, women recall carrying this family treasure, sometimes disassembled into portable parts, to their rural destinations of evacuation from firebombed cities.[97] Both wartime dress reforms, which expected and encouraged women to produce clothing at home, and the more spontaneous turn to *monpe,* resulted from and reinforced the already long hours of home sewing. Scarcity turned women toward home sewing on other home fronts as well. But women elsewhere were drawn into the wartime work force earlier and in greater numbers and proportions than in Japan; with wages to spend and little time to sew, they turned more than ever to the purchase of ready-made dress.[98] This did not happen in Japan. To the war's end, by machine or by hand, women sewed for themselves and their families. By the awful summer of 1945, sewing had sunk deeper roots than in the past as part of their identity as women and as managers of the home front and home life.

CHAPTER 6

Mechanical Phoenix

Among many stories marking the fiftieth anniversary of the end of World War II, the *Asahi* in August 1995 ran a feature on reporters who told the story of the atomic bomb. It focused on Kusakabe Hisajirō, the Nagasaki bureau chief of its competitor paper, the *Mainichi*. Kusakabe lived with his wife of one month, Toshiko, in an apartment behind the paper's office, sufficiently distant from the epicenter for the two of them to survive by diving into their air raid shelter just as the blast blew the roof off the office. Fifty years later, Kusakabe vividly recalled his frustration at the censorship that prevented him and his colleagues from publishing a detailed account of the damage and suffering in the city. He also recalled venturing out with his wife to look for her mother, who lived closer to the epicenter. Amid the destruction, they could not tell exactly where her house had been. "Guessing 'maybe here,' we started digging through the rubble, and soon came upon the familiar Singer sewing machine. Pushing further through the ashes of the *tatami* (floor matting) I found a finger with a gold ring on it. My wife said, 'This is mother's ring.'" Kusakabe gathered the burnt remains and put them into a metal helmet.[1]

The sewing machine does not figure this prominently or wrenchingly in every recollection of war survival and loss. But neither is its appearance surprising. The sewing machine was already an emblem of middle-class family life in the interwar era and wartime. By the 1960s, it would find its way into the great majority of all Japanese homes. But in sharp contrast to prewar times, Singer machines accounted for a minor share of postwar purchases. This chapter explores the process by which Japanese producers came to dominate home and world markets, and Singer's frustrated attempt to regain its dominant position.

The story carries forward two themes from the prewar era. One is the nationalized, at times racialized, understanding of business competition, already prominent in the clash between emerging Japanese producers and Singer in the 1930s. Renewed competition in the 1950s recreated a politically volatile structure of trade friction that would remain in place for decades. Japanese government and business strove to exclude foreign investment. American manufacturers leveled charges of "social dumping" and unfair competition. Sentiments expressed by *mishin* producers and promoters of "buy Japanese" campaigns of the 1930s resurfaced in the 1950s in powerfully emotional appeals to defend "ethnic industry" against American invaders. "Building a nation of culture and peace" had become the watchword of the day, but the rhetoric of struggle against an American behemoth was harsh and warlike.

A second continuing theme is the construction of modernity as globally connected but locally inflected. In the 1930s, Singer's "universal" practices had been resisted by the company's sales force and modified by Japanese competitors. Salesmen had criticized their terms of employment as inappropriate for Japan. Competitors had targeted a broader customer base with what they called Japanese-style credit. Even though similar practices existed elsewhere, and even as they marketed their machines as "just like the imported brands" and as a way to take part in a global modern life, both the Singer salesmen and the Japanese producers flourished rhetoric of "appropriately Japanese" practice. This complex dance continued in the postwar years of recovery, high growth, and mass consumption. As domestic *mishin* companies recovered and thrived, they developed a distinctive system of production but closely followed the Singer system of selling. Ironies abounded. These same firms abandoned much of the Singer system in their export strategies. They turned to alliances in the United States with mass discount chains, even as Singer stuck to its famed system both in America and in its effort to reenter Japanese markets. Yet at home, both the production and the selling practices of domestic firms came to be seen as particular to Japan. They described their efforts with a discourse of indigenous innovation—and resistance—even as they replicated much of Singer's familiar "universal" model: at the heart of their efforts were the face-to-face appeals of thousands of sellers and the enabling, but disciplining, power of installment credit.

STARTING OVER

Mrs. Kusakabe's family sewing machine was one among many thousands ruined in the rain of fire and atomic bombing in the war's final year. The Japan Sewing Machine Industry Association estimated that more than half of the nation's stock of household machines (800,000 out of about 1.4 million) had been destroyed in 1944 and 1945; as a result, the ownership rate fell from nearly one in ten households to one in every twenty-four homes.[2] From this devastated landscape, the sewing machine emerged as both mechanical chameleon and phoenix.

As chameleon, whether in marketing slogans, in the view of the Japanese government and American occupation authorities, or in the actual uses to which it was put, the *mishin* in the autumn of 1945 quickly changed colors from home-front weapon to implement of peace. As phoenix, it rose with astonishing speed to play an important role in the recovery of Japan's machine industry in export and domestic markets. Already in 1947, Japanese companies manufactured 134,000 household machines, followed by 166,000 in 1948, making it one of the first industries to surpass its peak of prewar or wartime output. Production of household machines as counted by the industry association approached 500,000 units by 1950. Output doubled the following year to pass 1 million, reached 2.7 million by 1960, and peaked at 4.3 million in 1969. In 1955, the value of sewing machine exports, at $38 million, was the greatest of any good in the light machinery category; it easily exceeded the combined export sales of the next four goods on the list: cameras, radios, bicycles, and binoculars. By 1957, 1 million out of 1.6 million sewing machines imported into the United States were made in Japan. As *Fortune* put it in 1959, "mighty Singer" was struggling to defend itself in the face of "the really alarming competition . . . of some 300 Japanese companies." Japan had become the world's leading producer of household sewing machines.[3]

The majority of this output was sold overseas; official statistics report that nearly 70 percent of the household machines made in Japan from 1948 through 1960 were exported. But, as government officials themselves recognized, their data significantly undercounted domestic sales in the first postwar decade. Exports were precisely tabulated by the Finance Ministry, but machines sold on the domestic market by the numerous large and small producers were not fully counted. A survey by the Ministry of International Trade and Industry (MITI) estimated that official statistics on sewing machine production missed nearly half of output in 1947–48, 30 percent in 1948–1950, and 20 percent in 1951–53. Revising the numbers accordingly, the ministry estimated that annual domestic sales of sewing machines, primarily household machines, exceeded four hundred thousand by 1950 and passed eight hundred thousand by 1953.[4]

These adjusted totals are certainly more accurate than the official count. If one accepts the latter, cumulative domestic sales by 1960 stood at just 5.6 million sewing machines, far short of the total needed for the sales figure to correspond with several convincing surveys of household diffusion rates. These make it clear that by the end of the 1950s, the output of Japan's producers in the postwar years combined with the existing stock of prewar Singer and Japanese-made machines—intact or repaired—had been enough to place a sewing machine in nearly three-fourths of Japanese homes. An Economic Planning Agency (EPA) survey conducted in 1956–57 and published in Janome's customer magazine found a sewing machine in 75.5 percent of urban households. By another count, in 1960, 72 percent of all 20 million households nationwide possessed a sewing machine, making it the second most widely owned consumer machine after the radio (89 percent) and well ahead of the television (55 percent).[5] Translating these percentages into numbers of machines, cumulative postwar sales and surviving prewar stock should have approached 12 million units, a total that roughly corresponds with MITI's adjusted estimates.

Janome titled its graph of the EPA survey "Ownership status of high-class cultural household goods" (figure 26). Such wording, along with the company's playful graphics in the style of Kon Wajirō, nicely reflects the excitement of the domestic producers as their fortunes soared through the 1950s. However, household penetration slowed some in the 1960s; it took until 1967 for the sewing machine to find its way into 80 percent of homes. Like refrigerators or washing machines, but unlike TVs or radios, this was a good generally limited to one per household. New models such as the "zigzag" machine, introduced in Japan in the 1950s, encouraged trade-ins and upgrades, but the golden age of household sales for the domestic market lasted only about two decades.[6]

The great irony in this impressive record of manufacture and sales is evident in the record of the Singer Sewing Machine Corporation. Not for lack of interest or effort, during the postwar decades Singer never won more than 15 percent of the Japanese market. It lost to Japanese competition around the world as well. But if Singer failed to regain even a shadow of its former dominance in Japan, its machine and its selling

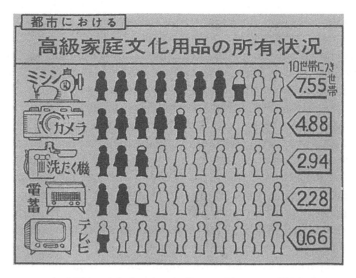

FIGURE 26. This graph showing the "ownership status of high-class cultural household goods" appeared in Janome Sewing Machine company's PR magazine, *Janome mōdo*, Spring 1958, p. 4. Its iconography mirrors the surveys and graphs pioneered in the interwar era by Kon Wajirō, and its title echoes an interwar vocabulary in which "cultural life" meant a life making full use of modern consumer goods. (Janome Sewing Machine Company)

system still cast a large shadow. Its effort to reenter the Japanese market and defend its place at home constituted an early episode in several decades of trade conflict that echoed prewar disputes and introduced the new dimension of free-trade anti-communism. Its selling system in Japan served as a model not only for the sewing machine industry but also for a wide range of consumer goods that remade everyday life and fostered a powerful new version of a demand-driven political economy. In this system's marketing appeals and selling strategies, in particular in the offer of consumer credit, women held a proud place as happy consumers and prudent home managers.

EMBATTLED PRODUCERS: SINGER VERSUS JAPAN

The points of origin for Japan's postwar sewing machine industry were three: a small number of manufacturers already established in the 1930s, a larger number of new entrants, and several "conversion" manufacturers, such as Tokyo Heavy Machinery (reborn as Juki). This last group shifted from producing weapons and precision machinery for the military to the making of *mishin;* in a peacetime economy, turning their production technology toward sewing machines was a logical step. The small-scale, network-based organization of production in this fast-growing industry differed from Singer's, which made most of its parts in-house. Half the companies surveyed by the industry association in 1956 were capitalized at less than ¥3 million (roughly $8,300). Firms with more than three hundred employees accounted for 40 percent of all machines produced, but just 20–30 percent of exports, where the small makers were stronger. The industry by the late 1950s consisted of fifteen major producers of finished machines with assembly plants. They subcontracted and purchased parts from a larger network of at least sixty-five principal suppliers.[7]

Organized in this fashion, Japan's sewing machine industry constituted a network-based system of part suppliers and assembly finishers. Constantly preparing for the day when "mighty Singer" might return to Japan, the goal of these producers was to match Singer's quality. Faced with the need to standardize machine parts produced by multiple firms, the industry took Singer machines as its models, provoking strong criticism from Singer for copying its parts, using its model numbers, and infringing on its trademark. Japanese producers ignored these complaints, and in 1949 they set Singer-derived industry standards (JIS, Japanese Industrial Standards) for their products. Noting their cost-effectiveness and flexibility in allowing the end-producers to adapt to changing market conditions, Kuwahara Tetsuya has attributed the success of the Japanese firms to this networked "assembly system."[8] Other scholars have made similar arguments for a range of industries, automobiles most famous among them.[9] The *mishin* producers were among the pioneers in creating what has come to be enshrined as a Japanese innovation in mass production.

As early as 1950 in the case of the Brothers company (at the time named Japan Sewing Machine) and 1956 in the case of Pine Sewing Machine, Japanese executives or engineers paid visits to the famed Singer factory in Elizabethport, New Jersey. They were astonished to discover that Singer no longer had much to teach them. These were among the first in a string of revelatory American sojourns by visiting Japanese engineers from all manner of industries from the early 1950s into the 1960s, although the revelation for others in the early years of this traffic was usually that of a huge technology gap.[10] One of the Yasui brothers recalled of his 1950 visit that "my host, the American factory director, guided me through the armhead, casting, parts, and assembly plants, but to be honest, in contrast to the immensity of the building, there was not much to surprise me about the facilities inside." The Pine engineer, whose company was eventually absorbed by Singer as part of its reentry strategy in Japan, recalled that "I gained a lot of confidence in discovering that in the previous ten years our own company's technology division had developed to the level of Singer with its history of more than 100 years."[11]

Although it organized production differently from Singer, and to good effect, in both product design and in sales strategy Japan's domestic sewing machine industry closely tracked the prewar giant's footsteps, listening all the while for the footfalls of a Singer return to the home market. Reading sources of the era, in particular the industry's lively trade paper, *Nihon mishin taimusu* (Japan sewing machine times), one senses that the retrospective claims by Yasui of Brothers (who wrote of his 1950 visit in 1961) or the Pine engineer (looking back at 1956 in a 1993 interview) were somewhat exaggerated. From its first issue in November of 1946, the pages of *Japan Sewing Machine Times* overflowed with complaints at the low quality of Japanese sewing machines, calls for industry-wide standardization of parts, defensive denials that Japanese manufacturers were stealing Singer technology and infringing its patents, and fearful reports that Singer was about to reenter the Japanese market.

Although Singer was indeed already interested in the Japanese market, the earliest postwar rumors of its return were premature, and the denials of infringement rang more than a little hollow. A small ad on the front page of the *Japan Sewing Machine Times* in March 1949 offers a glimpse of the problem with a "chart of all the new model numbers for parts set by the Ministry of Commerce," under the headline "The use of Singer model names or part numbers is prohibited." It wraps up the ad with the exhortation "Let's take care not to violate this."[12] Clearly, many manufacturers were not only copying Singer parts, but also using Singer part numbers and model names to signal their achievement. Despite exhortations to follow the new rules, from 1948 the *Japan Sewing Machine Times* itself was helping the cause of cloning in articles detailing the features of roughly two hundred Singer models already in use in Japan. From late 1949 through 1950 the paper published dozens of articles presenting diagrams and technical specifications of Singer machines and parts, along with their associated patents. The paper called this an effort to educate makers to respect patents, but it equally offered the possibility of making copies based on the diagrams.[13]

Singer was well aware of these endeavors. Alvin K. Aurell, the executive at New York corporate headquarters responsible for Asia and Africa, had served Singer as a manager in Japan from 1919 through 1939. On an ill-timed business trip to Manila in December, 1941, he ended up confined by the Japanese occupiers in a Philippine prison for two years. As early as November 1945, he prepared a detailed account of the company's war losses in Japan and its empire to initiate a claim for compensation. Aurell estimated the company had lost much more than $10 million in machine inventory and bank balances confiscated throughout Asia when the war broke out.[14] With an eye toward eventually reestablishing a presence in the Japanese market, Singer put its immediate postwar efforts into pursuing this claim, along with monitoring and protesting what it called illegal or unethical practices by Japanese competitors. It faced difficult going, not least because the American occupation authorities placed low priority on helping Singer or other United States businesses.[15] Even sending a company representative to Tokyo was a struggle. In October of 1947, Earle E. Lawton, another executive with prewar experience in Japan (1932–40), was allowed back as the firm's first postwar manager.[16] He had begun seeking permission fourteen months earlier by hand-delivering Aurell's letter on his behalf to the War Department in Washington. The missive pleaded that "it is highly important that an American representative [of Singer] be permitted to go to Japan" to investigate the fate of prewar bank deposits valued at $2 million, as well as machines and office buildings, including property in Korea; the status of "the many Singer patents and trademarks formerly registered in Japan"; and "the present competitive situation in Japan."[17]

Prior to Lawton's arrival, Aurell himself managed a brief tour in the spring of 1947 to Manila, Shanghai, and Tokyo to assess Singer's position and prospects. The Tokyo visit was hardly welcomed by American authorities. P. A. McDermott, chief of the Foreign Trade Division under SCAP (Supreme Commander for the Allied Powers), was back in Washington when Lawton spoke to him to help arrange Aurell's visit. McDermott "could not see where foreign business would be able to do anything for at least five years," and he warned Lawton that "for years the Japanese will have no yen with which to pay for anything but the bare essentials and he doubted if they would even have sufficient for this purpose." The latter, speaking from experience, disagreed: "Personally I cannot conceive of the Japanese not making a 'come-back' much sooner than he estimates."[18]

Aurell went to Japan not as a Singer representative but "in an advisory capacity to the Foreign Trade Division of SCAP's Economic and Scientific Section (ESS) for a period of two weeks."[19] He wrote home in April 1947 in great distress that the comeback was already underway: "I am very unhappy about business affairs here. It is an awful mess and I feel that S.C.A.P. are doing much harm to American enterprise. I shall not say much now but my mind is working overtime. The Japanese, with American support and their blessing, are going into sewing machines in a big way. To my mind something must be done about the present policy."[20]

In a slightly more measured tone, Aurell's formal report to Singer's eminent president, Sir Douglas Alexander, gave some sense of the gray zone the company faced with regard to competition that was not necessarily illegal, as well as the conceptual challenge that he and his colleagues would face in taking this competition seriously. Noting that, compared to "four or five sewing machine manufacturers in Japan before the war," there were now "23 manufacturers registered with the Foreign Trade Division of ESS" producing 2,500–5,000 machines per month, he told Alexander that "while these Japanese imitation machines may not infringe basic patents which have long since expired, they are nevertheless imitations of inferior quality of an original American product. They are made and advertised as having interchangeable parts with Singer machines.... It is unethical in every moral sense." Aurell on occasion turned this anger into a dismissive view of Japanese producers. Learning that industry representatives had told ESS officials that "by 1940 local production was at the rate of 200,000 per annum," he told Alexander, "This statement I can brand from personal experience as false. The obvious reason for making it was to convince the Americans that in being allowed to resume production now they are only taking up where they left off when the war began." The Japanese claim, however, appears to have been exaggeration more than outright falsehood; the industry association in 1940 had reported total production of 156,000 machines.[21] Aurell's "personal experience" in Japan ended in 1939, but one imagines his subsequent detention in the Philippines left a bitter legacy. He may have been unaware of the 1940–41 surge in local production, but he also seems unable to imagine it had taken place. Whatever his view of the past, he was clearly worried about the future. He concluded his letter by expanding his concern to American business more generally: "Japanese manufacturers are offering imitation automobile repair parts for export. Japanese cameras are now offered for export. Perhaps this is all legitimate and perhaps there is nothing that can be done about it but do General Motors, Ford, Chrysler, Eastman, General Electric and others know about it? It is certain we as an American company did not know the things I have learned until I visited Japan."[22]

Even for Singer, the production of knock-off parts and machines based on expired patents was not so much illegal as "unethical." Aurell complained with greater legal bite about trademark violations, especially "the flagrant misuse of our Singer transfer-marks (decalcomania). . . . It will be recalled that before the war it was a constant fight to control such infringements in Japan. While in Tokyo I personally saw evidence of Japanese imitation head castings and old Singer head castings being redecorated by the use of imitation Singer transfer-marks," which were said to double the selling price of a Japanese made machine. Aurell concluded in a subsequent letter to a fellow executive that "the Japanese were always imitators and always will be."[23]

Lawton received U.S. government permission to enter Japan in the fall of 1947. Nominally his position, like Aurell's, was not that of a Singer manager. He was a technical advisor attached to the United States Reparations and Restitution Delegation. In fact, he spent much of his time on Singer business. He rehired as local Singer agent the company's final and most loyal prewar employee, Kimura Kenkichi, who had reluctantly overseen the disposal of Singer machines after the military confiscated them on the day of the Pearl Harbor attack. The two set up an office in Tokyo with instructions from Aurell to "clarif[y] our war loss claim, looted machines, etc." and report "on the general situation and in regard to competitive developments, M.T.D. prospects, Patent and Trade Mark questions, etc."[24] From January 1948 through 1950, he sent a steady stream of complaints to various occupation offices about misleading advertisements, such as the Rhythm company's ad for its "Singer Type 15K83" machine; infringements on Singer patents that had not yet expired; and numerous trademark violations. One major claim focused on the use of the distinctive Singer oval mark by thirty-five or more Japanese machine and part manufacturers. One imagines Mr. Kimura making the rounds of sewing machine shops to uncover these violations, while Lawton followed with furious missives to SCAP condemning what he called the "unethical practice" of each "unscrupulous manufacturer" of an "inferior Japanese made product."[25]

Occupation officials took care to distinguish the ethically dubious from the illegal; they dismissed the complaint that Japanese machines were labeled as "Singer type." Such actions, they told Lawton, did "not represent their goods as those of any other manufacturer," and indeed directed the purchaser's attention to "the fact that there is a similar article but of different origin."[26] In other cases, Lawton and his successor as head of Singer operations in Japan from 1950, E. L. Jones, enjoyed nominal success. In response to Singer complaints relayed by the occupation authorities, the Ministry of Finance in 1950 judged nearly sixty sewing machine and part producers to be guilty of "Infringement of Trademark or Design of the Singer Manufacturing Company."[27] And as seen in the notice placed in *Japan Sewing Machine Times* in 1949, Singer was able to press the Ministry of Commerce to defend the cause of trademark and design protections.

Singer also succeeded in winning compensation of slightly under $1 million (¥339 million) for the eight thousand machines, along with parts and needles, that were confiscated by the Japanese military from its Yokohama warehouse on December 8, 1941, although its claim to $2 million worth of prewar yen bank deposits was denied. The Ministry of Finance settled this claim in 1953 with a payment to Singer from a government fund set aside for "war loss" compensation by order of SCAP. Along with IBM and Kodak, Singer's settlement was among the first. The amount was fourth largest of the fifteen major cases settled between 1953 and 1960.[28]

To the chagrin of Singer, Ministry of Finance officials had initially proposed locating as many of the machines as possible, taking them back from the individuals or companies to which the military had sold them, and returning them to Singer. Told that the Finance Ministry had located one thousand of the machines, Singer's Tokyo manager—who of course wanted cash money rather than a cache of old machines—wrote to Lawton, now back in New York, that "Mr. Kimura, who has good connections with the Finance Ministry and the Mitsubishi Trust Company who are handling the matter of restoration" tried his best "to persuade them to give up the plan to restore such machines."[29]

It turned out that only 361 of the machines were located intact; the ministry deducted their value from the compensation payment and authorized Singer to reclaim these machines. Singer first considered sending "each of these customers a letter from the Company explaining the circumstances and stating that instead of taking away their machine we would accept a nominal price of say 3,000 yen," but concluded it would be better simply to allow the "found" machines to remain with their owners. After the Japanese government "wrote individual letters to these people advising them of our decision," Singer's Tokyo manager, noting that the possessors included a number of regular customers for industrial machines, wrote New York that "we have seized the opportunity to promote further good will by supplying Tokyo and Osaka supervisions with a list of the customers having these restoration machines in each of their territories." For good measure, he passed on a letter of gratitude from one such customer.[30]

Despite this effort to cultivate its market one customer at a time, Lawton and his successors were less successful in selling new machines than in winning compensation for old ones. Lawton's sales efforts through 1949 focused solely on placing Singer machines in the PX stores for occupation personnel.[31] He and his successors then gradually turned to rebuilding a sales and service network for Japanese customers, but their results were meager. From early in 1950, the company began importing new machines for industrial customers.[32] For this year and 1951, when it was selling only to garment makers and other industrial customers, no sales figures for Singer alone are available, but total sewing machine imports to Japan in these two years stood at just 448 and 3,213 units respectively. Even assuming (reasonably) that Singer took the lion's share, these were tiny numbers.[33]

In 1952 Singer returned to the household market. The *Japan Sewing Machine Times* opened a front-page editorial that June in a foreboding tone: "It seems 'the inevitable has finally happened.' Singer is starting to sell household machines."[34] The previous week the paper had reported that recent rumors were correct: Singer had reconstituted a portion of its old network of shop managers, contracting with sixty stores to serve as authorized (not necessarily exclusive) distributors. In Japan, Singer's shift from exclusive reliance on company shops to selling as well through authorized dealers had begun in the 1930s. It appears that in 1952, Singer lacked a sufficient local presence to support its own shops. In a sign of further flexibility, Singer was also reportedly going to place its machines alongside domestic makes at major department stores, perhaps even the famed Mitsukoshi. The reaction of domestic producers was said to be appropriately calm for the short run. Singer would be importing at most one thousand machines monthly at a time when Japanese producers were selling at least forty times that number. With a price-tag predicted to be ¥60,000 for Singer's most popular household model—triple the cost of a comparable domestic make—sales were expected to be modest.[35] But over the long run, both the newspaper and its industry readers were worried: once Singer reconstituted its famed selling system and secured a foothold in department stores, the combination of brand reputation, product quality, comprehensive service, and installment credit meant that monthly sales might rise from one thousand to ten thousand or more. Without ceaseless efforts to raise quality and strengthen its selling system, "the possibility is great that the [domestic] industry might return to its prewar position."[36]

In the event, these fears proved groundless. Singer sold 2,784 household machines in 1952. Sales rose to 6,725 in 1953, 6,872 in 1954, and 9,308 in 1955.[37] Not until late in 1954 did Singer begin to offer installment credit to household customers.[38] Whether compared to Singer's Asia-wide annual sales of 130,000–150,000 machines, or compared to a total Japanese market estimated to be 800,000 machines in 1953, these sales totals were negligible.

The problem, it seems clear, lay in the double punch of Singer's high price and the growing quality of the machines made in Japan. Singer soon decided that the only way to confront its new competition, not only in Japan but in other Asian markets, would be to produce in Japan. Reports began to appear in the industry press in spring of 1953 that Singer was looking for a partner to undertake local production.[39] Fourteen months later, the rumor mill proved right. In July 1954, A. K. Aurell came to Japan with Singer's young new president, Donald Kurchir, and the company announced its plan to take a 50 percent ownership stake in Pine Sewing Machine Company, a struggling subsidiary of Mitsui-connected Japan Steel Corporation with a factory in Utsunomiya, northwest of Tokyo.[40] In addition, Singer proposed to share its technology with Pine and invest ¥100 million (about $280,000) in new production equipment for the Utsunomiya plant. Its initial goal was to produce five thousand machines per month, about 5 percent of total Japanese production at the time.[41]

A firestorm of patriotic protest erupted. Immediately after the announcement, the *Japan Sewing Machine Times* came out with its first ever "extra" edition, one dozen pages of purple prose on the "S-P Question." In the following months, a powerful alliance of industry associations, manufacturers, and labor unions pressed the government to prohibit the investment. *Fortune* magazine called it "an anti-Singer campaign complete with torchlight parades, shrill banners, marching school children, and affecting tableaux."[42] The loathing for and, even more, the fear of Singer's invasion contrasts sharply with the recollections of men like Yasui of Brothers, whose recent visit to the United States left him confident that his company had caught up to Singer in technology and productive capacity. The tone was shrill indeed in the many statements in defense of this industry, which had, according to a July 19 declaration of the Sewing Machine Industry Association, "risen so swiftly from the postwar ruins as an export industry" to "reach global standards of quality. . . . This incursion of monopoly capital is completely unprecedented." Singer would use its capital and its powerful sales network to "pierce our domestic market and eat away our world market." The "self-reliant productive system anchored in hundreds of small businesses will collapse and the livelihoods of the 1.2 million Japanese who depend on this industry will be plundered."[43]

As the protests escalated, it became clear that MITI, relying on the 1950 Foreign Investment Law (FIL), would not grant Singer access to the foreign exchange needed for its proposed investment. The United States government protested what it called a violation of the spirit of the U.S.-Japan Treaty of Friendship, Commerce, and Navigation (FCN) of 1953, which guaranteed "most-favored nation" treatment to American companies.[44] But insofar as the more specific FIL, written and implemented under the watchful eyes of the American occupiers, trumped the FCN, MITI was in the driver's seat. Unable to invest dollars in the Pine Corporation, in late 1954 Singer and Pine executed a fallback strategy. Singer's Japanese sales office avoided the strictures of the FIL by borrowing yen locally and using its own yen assets, including a portion of the ¥339 million received in 1953 for its war loss claim, to take a 50 percent share in the company and upgrade Pine's technology. Fearing that the machines fell short of " 'Singer' quality and standards," Singer was at first unwilling to sell under its own brand.[45] In November 1955, "as an interim measure," the Pine factory began to produce about fifteen hundred "Merritt" sewing machines per month for Singer. The brand came from Isaac M. Singer's middle name. Singer sold these machines through a network of mostly non-exclusive outlets.[46] The company had precociously set up what the Japanese government would legally recognize in 1956 as a "yen-based company." The downside was clear: Singer could neither repatriate profits from Japan nor send dollars to invest in the Pine factory.[47]

This move, no less than the initial proposal, met fierce criticism. Janome's President Shimada penned a fervid call in the *Japan Sewing Machine Times,* to "defend ethnic industry." Probably referring to Singer's use of war-compensation funds in this deal, he claimed that Singer was financing its "deceitful invasion" with "none other than monies stolen from the pockets of Japanese people before the war."[48] Drawing on the military's slogan from the summer of 1945, one writer for a weekly magazine told readers that "the Singer problem has now turned into a final battle in defense of the homeland, and the key to victory or defeat is in the hands of the housewife." Yet, in the same paragraph, this writer noted that "Japan's major producers have improved their quality and technology, worked to lower their costs, and are now strong enough not to be defeated in competition with Singer."[49] Even the industry association in its doomsday declaration of July 1954 had boasted of having reached global standards of quality. The fear of Singer's invasion came from the gut as much as the mind.

Within a year of its investment, Singer reported that "much has been accomplished in the task of re-tooling the factory."[50] Satisfied as to the quality of the product, Singer shifted from the Merritt to the more powerful Singer brand name, increased the number of its own retail stores, and began to lure customers with its tried-and-true techniques of direct selling, "easy payments," and free lessons.[51] By 1959 Pine was producing more than seventy thousand machines for Singer annually at its Utsunomiya plant, and Singer was exporting ten thousand of these to other Asian countries.

Finally, in late 1959, the government gave Singer what the Japanese press termed "a present to American industry" on the occasion of a visit by Finance Minister Satō Eisaku to the United States: a promise of permission to proceed as originally proposed, investing dollars and taking home profits. The response at this point was relatively calm; over the previous five years, the industry had survived the competition with Merritt and then Singer machines. Some observers did, however, express fear that because Singer could more easily invest to upgrade Pine's technology, it would be able to expand output greatly for sales in Japan and for export, and advance in the "struggle to regain global hegemony."[52]

In the epilogue to his important book on racism in the Pacific War, John Dower asks, "After such a merciless war, how . . . could the race hates dissipate so quickly?" He answers that in addition to the simple, in-person discovery that their former enemies were neither demons nor brutes, the "stereotypes which fed superpatriotism and outright race hate were adaptable to cooperation."[53] As Dower contends, the malleability of America's simian image of Japan as it transformed from sneaky to cute, or of Japan's demonic image of Americans, shifting shape from destructive to protective, probably did contribute to reconciliation. But this interpretation overlooks the power of sublimated hatreds of wartime now displayed on economic and cultural grounds. The warlike and feverish nationalism that greeted the Singer-Pine proposal offers a glimpse of the enduring force of earlier fear and enmity. Consumers were told to be loyal and buy Japanese. Producers called on the state to protect "ethnic industry." Singer was described in ferocious tones as a predatory invader.

Although America's trade policy was considerably more liberal than Japan's, economic nationalism on the American side of the Pacific was similarly harsh. The United States maintained a considerable trade surplus with Japan in the 1950s, but currents of complaint at "Japan's traditional tendency to dump poor-quality products on world markets" ran strong and steady.[54] Leading these protests were spokesmen for industries ranging from plywood, tuna, chinaware, and textiles to cameras, watches, and sewing machines. The vocabulary of war was ubiquitous. In 1956 *U.S. News and World Report* wrote of Japan's "all-out attack on Western markets," and in 1959 it described an "import invasion," as American manufacturers "battled a rising tide of low-cost imports."[55] That same year *Time* magazine described Japanese competition as a "full-scale commercial assault."[56] And in 1958, *U.S. News and World Report* trotted out the granddaddy of war clichés, quoting an umbrella maker: "We are again being attacked, Pearl-Harbor fashion."[57]

The saga of Singer's return to Japan and Japan's *mishin* exports to the United States reveals at the level of discourse both ongoing elements and shifting patterns of trade-war talk. Echoes of the 1930s criticisms of cheap labor, unscrupulous mimicry, and social dumping, mixed in with the vocabulary of war, had been clear in the occupation-era statements of Singer's management. But one element in American trade talk was new: the call to accept Japanese imports in the interest of strengthening a cold war ally. The fear that Japan might suddenly "go red" was always present. As the *Saturday Evening Post* put it in 1952, "containment of Russia and its satellites can't be had without a price," that price being Japan's access to the American market.[58] In addition, a gradual shift came with recognition in some quarters of Japan that protection was no longer needed, and with admission on the American side that high Japanese quality as well as lower cost was part of the problem. Among Singer executives, the scornful assumption that Japanese manufacturers produced inferior goods through unscrupulous means had vanished by the time the company decided to make its product in Japan. Ahead of its competitors in this regard, Singer—aware that its future as well as its past lay in the ability to sell not just at home but worldwide—was also among the earliest firms to push more for reciprocal access in Japan than for protection at home.

Singer had been one of the few foreign firms with significant prewar reach into Japan's household consumer market, and it had maintained a presence, however defensive and diminished, until the Pacific War began. It was also unusual in making vigorous, early postwar efforts to sell in Japan and invest in local production, well before others saw Japan as a market worth fighting for. Some producers of industrial goods, most notably Texas Instruments and IBM, sought market access and direct investment alongside Singer in the early-to-mid 1950s. But among sellers to retail consumers, only Coca Cola tried as hard (facing similar frustrations) to enter the Japanese market by producing and distributing locally in these years. Among producers of mechanical or electronic goods and appliances, such as Eastman Kodak, GE, or Westinghouse, Singer's determination and its strategy of producing in Japan and selling through its own local network stood out.[59] It was a company with a unique and long-standing commitment to global markets.

Yet Singer never regained more than a fraction of its prewar market share. If the company correctly understood Japan's potential and made the "right" move in pursuing it aggressively, why were its results disappointing? The protectionism of Japanese producers and of MITI was only part of the story. After all, within eighteen months of proposing to invest in Pine, Singer was able to make an end run around MITI and begin selling Japanese-made machines in Japan—albeit at first under the Merritt label. It is hard to imagine that this delay sealed Singer's fate in Japan.

Rather, the success of Japanese producers in putting out high-quality products at low cost, combined with Singer's reluctance to compete aggressively in price, even after it was producing with the same local cost structure, seem to be the key factors in its relatively modest reentry. Singer did expand Japanese sales somewhat in the 1960s. It rebuilt its network of exclusive Singer stores, which coexisted uneasily with authorized retailers handling multiple brands. Production capacity at its Utsunomiya plant rose steadily to 324,000 machines per year in 1970, 7 percent of total production in Japan (4.35 million units). Singer exported about 40 percent of these machines to other Asian countries, below the 67 percent export ratio of all producers. The nearly 200,000 machines sold in Japan that year gave Singer a 14 percent share of the domestic market.[60] While not a dismal failure, this was far from its prewar dominance.

The company very likely could have expanded its reach in Japan by offering machines at lower prices. In Europe and North America it took this course from 1958 by introducing a low-priced Spartan model featuring "plain straight stitching . . . to counter the cheap Japanese machines which have made serious inroads."[61] But even as it did this, Singer gave greater priority to the high end of the Euro-American market, where it had slipped in the face of European and then Japanese makers of the "zigzag" machine, which was capable of a more flexible and stronger stitch. With its new "Slant-o-matic" machine, in the early 1960s Singer launched a drive to regain the lead in this market.[62] Some of Singer's Japanese managers apparently argued that the company should focus on selling Spartan-type machines in Japan.[63] Such pleas were ignored, and Singer began to manufacture the "Slant-o-matic" at the Utsunomiya plant in

1964.[64] But this machine cost an entry-level male office worker in Japan about three months' pay. It was too expensive to find a mass market.

Singer seems to have been half-hearted in all markets about the "Spartan" strategy. The company in 1960 admitted that in Asia "our merchandise is generally higher priced than that of our competitors," but went on to say that it was gaining back market share through "the excellent quality and reputation of our products, our willingness to extend credit, and the enthusiasm and determination of our sales organization."[65] This was the traditional Singer philosophy. As the head of the consumer products division in the United States said in *Forbes* in October 1964: "I intend to beat the hell out of the Japanese" through lower production costs coupled with more service and more deluxe machines with unique features.[66] In addition, Singer at the end of the 1950s, although lagging in Japan, was still running even with its Japanese competitors elsewhere in Asia. With this broader market in view, Singer faced the dilemma that selling more cheaply in Japan would have undercut its pricing elsewhere in Asia. It was also unwilling, apparently for political as much as economic reasons, to follow the lead of two American competitors by importing back into the United States machines made in its Japanese factories.[67] All these factors kept Singer focused in Japan on the high end of the home sewing market, and they almost certainly did impede sales. But given the gains of Japanese competitors in the 1930s and their postwar improvements in quality and productivity, no strategy would have allowed Singer to regain completely its prewar hegemony.

THE SCIENCE OF SELLING AND THE SELLING OF PLEASURE

Reborn as a strident producer of instruments of peace, the Japanese sewing machine industry in the early postwar years resisted the Singer "invasion," in large part by making and selling machines unabashedly (and accurately) marked as "Singer-type." A two-fold irony attended the worldwide surge of Japanese competitors juxtaposed to the modest extent of Singer's postwar return to Japan. In North America, Japanese firms eschewed the Singer system in their sales strategy, even as Singer clung to it: Singer's Japanese competitors won the day by placing low-cost machines in the chain and discount stores that were reshaping the landscape of American retail selling.[68] Yet in 1958 it was a considerable innovation that Singer in the United States even placed its accessories, such as threads and needles, in department and chain stores, and Singer executives were adamant that "this was not a prelude to putting Singer machines in variety stores." In 1959 Singer still sold its machines in the United States and Canada through an exclusive network of 2,068 shops staffed almost entirely by Singer employees.[69]

But in Japan, Singer's newly confident competitors stayed with a Singer-like selling system, tweaking and rebuilding it into what eventually became the model for all manner of consumer appliance and electronics manufacturers: an expensive selling network of local shops featuring door-to-door sales, demonstrations and home trials, installment plans, and strong service. Some of these firms, such as Matsushita, had been developing retail networks in the prewar era, and sellers in various fields were surely taking cues from each other as the domestic market surged.[70] But Singer preceded all of them, and it surely was an important reference point for others.

In the mid-1950s, the major sewing machine producers began to follow Singer's system of selling more closely than before. Step one was to create their own store networks. As domestic producers had begun to compete with Singer in the 1930s, only a few, most prominently Janome, opened company shops staffed by their own salesmen. But even Janome also sold through independent dealers that offered multiple brands. And most firms, including successful entrants such as Brothers, relied mainly or entirely on such dealers. In the early postwar years, the Japanese *mishin* makers had their hands full setting up efficient, high-quality production systems. They remained content to leave the selling to such dealers. Even Janome through 1954 maintained only thirty-four direct sales outlets. It sold three-fourths of its machines through independent dealers.

The proposed Singer-Pine venture sparked a change in strategy by creating a sense, in the words of future Janome president Shimada, "that the enemy's invasion fleet was floating just offshore." Janome resolved to quintuple its sales force and sell exclusively through company stores. By 1958, it ran 156 stores and had cut all ties with independent dealers. By 1960, it had added another 100 stores.[71] Others joined Janome in this new approach, at least in part. As Brothers told its postwar story in 1971, facing fierce competition among dozens of makers, the company could not rely on independent dealers to prominently display and promote its product. It had to "compete along with most of the other firms in setting up a direct sales system." Its reasoning came straight from Singer's playbook: the value of "creating a location for a dialogue directly with the customer, allowing us to provide sufficient after-service, and build a sense of comfort and familiarity with the customer" justified the high fixed cost of a direct sales network. Such logic notwithstanding, Brothers and other firms did not move entirely to sales through company stores. Brothers by the end of the 1960s maintained nearly six hundred company shops under eleven regional offices on the Singer-Janome model, but the company also sold through nearly a thousand authorized independent dealers.[72]

To make the expensive investment in company stores and a sales force pay dividends, the major *mishin* producers gave new attention to the science and system of selling. A Janome spokesman in 1965 noted that "eighty percent of the battle is having sellers call on the right people."[73] Since the 1930s, Janome had followed Singer in using "prospect cards" to guide salesmen to potential customers. But until the 1950s their use was limited; salesmen for the most part made inefficient cold calls door to door. The remedy hit upon by Janome and its competitors in the 1950s, and used by other appliance sellers, was the "street" exhibit. In the afternoon and evening in neighborhoods throughout the country, aiming to catch the attention of women on their daily round of food shopping and errands, the sellers set up shop on the street with their sewing (or washing) machines (figure 27). They offered demonstrations and the chance to try out a machine on the spot. Collecting the names and addresses of those who stopped by, the sales force generated a prospect list for home visits the next day. Janome believed these exhibits were a most efficient use of sellers' time. Each store was required to exhibit at least ten times monthly. The goal was to generate 60 percent of sales contacts through these events, which also served to train new sales people as they worked alongside veterans.[74]

As with its prepayment system, which it had presented in the 1930s as derived from native tradition, despite its similarity to American practice, Janome drew on a vocabulary of Japanese-ness to describe its decidedly modern, American-style selling system. As company publicists told the story, Janome's system focused on "intensive, deep cultivation" of an assigned territory by each salesman. The phrase played upon an agronomic expression for what Janome called "the typical Japanese method of agriculture." This method was said to be of particular importance for sewing machines, which required more intensive persuasion, education, and follow-up services—including payment collection—than most products.[75]

Janome's commitment to system and science in selling shows as well in the January 1956 founding of a publication for the company's thirteen hundred sales and service employees. In the founding issue, Presi-

FIGURE 27. This photograph appeared on the cover of the July 1969 edition of Janome Sewing Machine Company's in-house bulletin for its sales force. It nicely captures a common scene in neighborhoods all over Japan of the 1950s and 1960s, when salesmen worked to convince women en route to or from their grocery shopping of the virtues of Janome products. It also makes clear that the customers have shifted to Western dress for their daily wear. (Janome Sewing Machine Company)

dent Yamada summed up his managerial philosophy as "Sales First": the "salesmen are the most valuable asset in our company," and, after all, "99 percent of the presidents of the major companies in America started out in sales."[76] Published quarterly for the first two years, then every other month, at first 6–10 pages an issue, expanded to 24 pages in 1962, this bulletin offered a passionate mix of inspiration and education to sellers, collectors, and teachers. A round table of collectors offered tips on how to be disciplined and firm in dealing with customers.[77] A shop manager told a cautionary tale of tolerating sloppy reporting and accepting too many "phantom" sales (faked by sellers eager to build their records, even if the customer returned the machine after one or two months). Eventually he learned how to offer tough love to those sellers truly willing to work hard and play by the rules.[78] Three female teachers (of the five hundred employed by the company) told of their pride in bringing happiness and confidence to women who learned to master this machine.[79]

Letters from users were printed so that sellers could learn what practical benefits and pleasures customers found in their machines. And every issue featured one or two top sellers, including a small cadre of saleswomen, who told model stories of success.

The typical narrative began with fear and failure, which was overcome by hard work, an upbeat attitude, and ingenuity. In practice, many sellers proved unable to follow this formula. As thousands of salesmen and a few saleswomen flooded neighborhoods around the country, the magazine *Nihon shūhō* (Japan weekly) in 1964 zeroed in on "the strange relationship between the sewing machine industry and its salesmen."[80] From all companies, these sellers were estimated to number 50,000 nationwide. Tactics were not gentle. To meet quotas, sellers stole customers, even after they had contracted with a competitor, by promising even more up-to-date features and paying the cancellation fees themselves. Turnover was high. New hiring was constant. Riccar, famous for its aggressive tactics and total reliance on installment sales, employed 12,000 salesmen in 1964; Janome was said to have 6,500 going door-to-door, and Brothers was said to employ 5,000 sellers.

Although the mature industry at its peak from the 1950s into the 1960s consisted of a dozen major makers, a number of smaller assemblers, and dozens of parts suppliers, these three companies dominated the domestic household market. Janome and Riccar alone were said to account for 70 percent of domestic sales in the early 1960s. On average in 1964 one of their salesmen took home an ¥1,800 commission per sale and sold about seventeen machines a month, for an income, including bonuses, of ¥40,000–¥50,000. The fixed base salary was small, around ¥3,000–¥5,000 yen. In this world immediate results brought immediate rewards. A top seller could earn ¥100,000 a month and win promotion to shop manager in his twenties. But the same high flyer would be quickly demoted to sales if not successful as a manager.[81]

This account concluded by arguing that the sewing machine, now replete with multiple stitching options to enable all sorts of embroidery and accessorizing, had become a leisure good. "Refrigerators or television serve the whole family, children and husbands included. Husbands entertain themselves with pachinko or mahjong or golf. Women have only their sewing machines. But with the appearance of the 'zigzag model' this machine has gradually become leisure-ized. Women are more and more pedaling the *mishin* not for a living but for the pleasure of making things."[82]

The words and images of sewing machine advertisers anticipated this shift by as much as a decade. From early in the 1950s, the *mishin* mar-

FIGURE 28. These three advertisements appeared in *Nihon mishin taimusu* in the early 1950s. From left to right, the ads are for Pine Sewing Machine (June 21, 1951), Kōyō Sewing Machine (August 15, 1952), and Mitsubishi Sewing Machine (January 1, 1954). This ensemble offers a view of recurring themes in the marketing of the sewing machine: sexualized glamour, modern science, and the happiness of the maternal fabricator of the family wardrobe. (*Nihon mishin taimusu*)

keters soft-pedaled the image of the diligent homeworker—who in fact remained an important customer—as they promised a good life to the happy owner of a home sewing machine. The promise of glamour and pleasure for oneself and one's family, and the invocation of scientific progress expanded on tried-and-true themes found in Janome's campaigns of the 1930s as well as Singer's prewar brochures (figure 28). But if little was new, the images were newly exuberant, whether in the sexuality of the Pine ad in 1951 or, perhaps most astonishingly, in the image of an "atomic" sewing machine used by Kōyō Sewing Machine for an ad that ran on August 15, 1952, the anniversary of the surrender. The decision to run this ad suggests that atomic imagery in Japan of 1952 evoked a sufficiently positive view of science to outweigh the association with the trauma of the A-bomb.

In addition to its bulletin for employees, Janome in the mid-1950s also launched a magazine aimed at customers, *Janome mōdo* (Janome mode). Issues from the 1950s and early 1960s tilted toward the sewer-as-producer. Stories of homeworkers as happy and successful *mishin* users were frequent. Typical was the tale of Kawaguchi Kinuko, who, by "simply doing Western sewing homework, has managed to save the sum of 1 million yen! So, those of you who are thinking of giving up because you don't have capital or much skill, wait a minute." A two-page spread

蛇の目社内報

第 5 巻第 24 号
昭和 35 年 1 月

蛇の目ミシン工業株式会社

FIGURE 29. This January 1960 cover illustration from the Janome magazine for its sales staff creatively links recent traditions of Meiji-era sewing, the ancient tradition of a form of badminton played in the Heian court (the illustrations appear on badminton paddles), and the happiness of a modern bride with her new sewing machine. (Janome Sewing Machine Company)

★去る7月米国ロングビーチで開か
れた1960年度ミス・ユニバースコン
テストに世界各国代表78名の中から
見事日本代表の児島明子さんが世界
第一位の栄冠を獲得しました。
★日本がこのコンテストに参加して
から八年目……日本中が待ち望んで
いた"美の女王"が誕生したのです。
★"世界一の美女"となつた児島明
子さんを、日本代表として送ること
に協力した蛇の目ミシンでは、この
入賞の報に社をあげて拍手を贈りま
した……というのは、児島さんが蛇
の目ミシンの熱心な愛用者でもある
からです。

世界一の美女児島明子さん
1960年ミスユニバース第一位の栄冠！

"世界一の美女"というお土
産をもつて羽田に下り立つ、
児島さんの晴れ姿。報道陣の
フラッシュに浮き出した児島
さんの瞳はうるんでいます。

FIGURE 30. The headline to this article from the Fall 1960 issue of *Janome mōdo* reads, "The World's Most Beautiful Woman." Kojima Akiko became a national heroine when she was crowned Miss Universe in 1960. Janome had sponsored her as a contestant and now proudly featured her endorsement as an "ardent and loyal user." (Janome Sewing Machine Company)

narrated in detail the five-year journey to success of this resident of Tokyo's plebian "low city." Her husband worked in a factory making electric machinery. She had worked in a dress shop before the war, and now, with a small child at home, she wanted to try home-based, commercial sewing. He is at first opposed: she is already tired by the end of the day from housework and childcare. But she loves to sew. She makes blouses for women in the neighborhood, turning a nice profit. When her husband recognizes her earning power, he brings home shirt orders from workmates. As her business grows, one of the greatest challenges is "keeping up with newest techniques of Western sewing." She reads dress magazines and newspapers, and clips articles on "popular trends" or the "connection between design and personality." After five years, she takes in ¥37,000 a month and clears ¥220,000 a year in profit. Her husband is so impressed he has started to come home early to help out with the housework. Her aim now is to open a store.[83]

Even this story of a producing consumer stresses the pleasure of the work for the ambitious Kinuko and her happy engagement with the world of design and fashion. And the prominence of commercial users like Kinuko gradually receded in Janome's marketing. The first color illustration ever to appear in the in-house magazine graced the January 1960 cover. It brilliantly combined an image of ancient tradition—the New Year paddle used in the Heian-era game of shuttlecock—with a Meiji-era tradition of a modernizing lady at her machine and a postwar image of a bride in a Western wedding gown (figure 29). Later in 1960 the company powerfully connected the sewing machine to visions of celebrity, glamour, and beauty with the endorsement of Kojima Akiko, Japan's first winner of the Miss Universe contest (figure 30). In 1962 Janome launched a "lovers' sale," an ad campaign that positioned the *mishin* as the key to a blissful romantic life. As *Janome Mode* and the in-house magazine became increasingly colorful and glossy in the 1960s, they illustrate a shift that took place in the message Janome projected to customers—away from the sewing machine as an economizing tool of the wage-earning, producing wife and toward the *mishin* as the indispensable tool of the happy bride and an instrument allowing a female consumer to fabricate a fashionable, good life.

THE HAPPY LIFE OF DISCIPLINED BORROWERS

As the most successful sewing machine makers imposed a marketing system on their salesmen and collectors, so they sought to discipline, as well as please, their customers. With the economy in high gear, the *mishin* sales force drew millions of buyers ever more fully into a life of rational planning overlaid with the pursuit of happiness. Installment credit was at the center of this effort. The purchase of sewing machines "on time" helped pave the way for a dramatic expansion of consumer credit as one engine powering the postwar economy.

Industry publications mention installment plans as early as 1948; credit was said to be in great demand by individuals, schools, and home economics teachers on behalf of schools. One story told of a group of policemen who jointly signed an installment contract to buy sewing machines, presumably for their wives. It was not unusual for a group of buyers to collectively guarantee each other's purchases in order to establish trust with reluctant installment sellers.[84] But most sewing machine makers, fearful that the early days of recovery would not last, were initially uneasy at aggressively extending credit. In 1949, competitors criticized Riccar when it first offered what it touted as "easy, American style" terms—an installment contract with no money down and no prepayments. With a reputation as a lone wolf, Riccar refused to join the industrial association.[85] Yet by 1950 sewing machine producers, together with makers of bicycles, motorbikes, refrigerators, washing machines, autos, and farm machines, were speaking of installment sales as the key to "healthy" and "rational" growth; they called on MITI, without success, for tax preference and a state-funded line of credit for sellers of approved installment goods.[86] Janome resumed layaway and installment selling in 1953 to good effect: sales overshot projections by 10 percent.[87] Juki entered a partnership in 1953 with Tokyo Mutual Bank as its finance company. To protect itself against default, the bank kept the customer's down payment in escrow until one-fourth of the installment payments were made, at which point it turned the down payment over to Juki.[88]

With sewing machines among the leading goods, by the end of the 1950s the installment plan had become the favored choice for women and their husbands and families who were eager to own the many "cultural goods" that defined the bright, new consumer life of postwar and peacetime. Their collective shopping binge transformed sewing machines and radios, then washing machines, televisions, refrigerators, vacuum cleaners, cameras, motorbikes, and (later) automobiles from luxuries to virtual necessities of the burgeoning middle-class. One 1959 survey enumerated the proportion of goods bought on monthly payments: bicycles, 80 percent; televisions, 75 percent; automobiles, 70 percent; motorbikes, 68 percent; refrigerators, 66 percent; sewing machines, 60 percent; washing machines, 59 percent. A 1960–61 MITI survey of 6,200 retail sellers offering a broader spectrum of goods, including clothing, found that from one-half to two-thirds of all sales at these stores were made through installment plans.[89]

The popularity of such consumer loans stimulated retail stores and non-bank lenders to offer a new and potentially far-reaching form of credit—a harbinger of the credit card. Called "ticket" or "coupon" financing, this form of lending emerged to prominence and political controversy in the 1950s.[90] It stood midway between a small loan to be used at consumer discretion and an installment loan tied to the purchase of a particular good. As with a credit card company, the ticket company contracted on the one hand with its affiliated retail stores and on the other hand with consumers. The latter were given books of yen-denominated coupons that they used in lieu of cash to purchase goods at member stores. The store would redeem the coupon for immediate payment from the ticket company. The buyer would repay the ticket company with installments spread over three to twelve months.[91]

Two vigorously competing types of retail sellers offered ticket credit: the small neighborhood stores clustered in the many shopping districts in all major cities, and their archrivals, the department stores. Nihon shinpan (literally, "Japan Credit Sales") was the pioneer in this field. To this day one of Japan's largest non-bank lenders, it was founded in 1951 in partnership with Tokyo's major department stores. In 1956, when it announced plans to branch out to Nagoya and Osaka, federations of small-scale retailers and two hundred allied ticket companies nationwide launched a campaign in opposition.[92] Their intensive lobbying led MITI in October 1959 to issue a little-known but highly consequential "Order Concerning Self-Restraint by Department Stores." This prohibited the nation's 106 major department stores from selling installment goods priced below ¥500 (¥1,000 in the six largest cities). It banned department-store installment sales of services, food, drink, plants, or pets. It outlawed the use of coupons outside the prefecture of issue.[93] Two years later, an Installment Sales Law—also drafted by MITI—reinforced these administrative constraints. It stipulated that "if it is judged that a department store or manufacturer will cause great harm to the business of small scale installment sellers, the relevant ministry can refuse to allow the former [to sell on installment]," and it directed the government to draw up a list of the "specified goods" that could be sold on installment credit.[94] The initial list of approved goods was extensive. It allowed credit purchase of items such as toys and cosmetics—hardly "consumer durables." But it prohibited "consumptive" credit—food, drink, and tobacco—as well as credit for services.[95] The small retailers had won a monopoly on installment credit for low-priced goods, and nationally portable consumer credit had been ruled out.

The direct impetus for the 1959 order and the subsequent law was a political struggle between large and small retailers, not fear of excessive borrowing or consuming in and of itself. The impact of these measures on credit behavior was significant; they channeled consumer credit toward the installment purchase of relatively expensive consumer goods, one contract at a time. These measures also shaped the understanding of consumer credit by entrenching in law a distinction between good and bad, or productive and consumptive, credit. They open a revealing window onto the widely shared ambivalence at the tight connection between individual borrowing, family consumption, and national prosperity.

Such concern was not new. A 1951 editorial column in *Japan Sewing Machine Times* had described installment purchasing as a benefit for the "national economy" but immediately reminded readers that it was dangerous in excess: "The installment buyer, when [this method] is made too convenient, tends to wastefully purchase items of little necessity."[96] As in the 1920s and 1930s, the late 1950s debate over consumer credit sparked much anxious commentary, although the hand-wringing in Japan was no greater, and perhaps less hysterical, than the denunciations presented by Lendol Calder as typical of American critics of consumer borrowing, and it was no more censorious than European consumer consciousness as described by Rosa-Maria Gelpi. In Japan the predicted damage fell into one of several categories of "excess": excessive pressure on consumers by unscrupulous sellers; excessive purchases plunging consumers deep into debt; excessive consumption sparking runaway inflation; and excess competition ruining the credit providers or retail sellers.[97] When observers stressed benefits, they pointed not only, or mainly, to easier lives for individuals, but to gains for the national economy.

The editors of a newly founded monthly journal of Japan's Credit Industry Association, *Geppu kenkyū* (Installment studies), prefaced a series penned by Satō Sadakatsu, a "counselor" in the government's Small and Medium Size Business Agency, with the claim that installment selling held the potential to "raise the cultural standard of living of ordinary people, a truly revolutionary technique for the world of commerce." But Satō himself raised the caution flag: only expensive products, not cheap necessities like food and drink, should be sold by installment. He lamented that the marvelous convenience of the coupons led many individuals to go overboard, and he worried that on the national level, credit was inflationary if its terms were too easy.[98]

The mainstream press was similarly ambivalent. In a feature of November 1957, "Advent of the Installment Age?" the *Asahi* noted that "all sorts of goods are selling this way: TVs, sewing machines, refrigerators, washing machines" but worried over reports "that installment is more expensive [than buying with cash], and people overspend." Although convenient for salaried workers, installment purchases not only risked ruining the household's finances if overdone; they could harm the "national kitchen." The writer decried the use of coupons for movies or in coffee shops. He also worried that while banks and the Finance Ministry could easily regulate the flow of capital to industry to slow down an overheated economy, they could not so easily regulate the more decentralized providers of consumer credit to head off inflation.[99] A vice president of the Japan Productivity Center—an important organization promoting greater efficiency as the key to higher growth—recognized in 1958 that consumer credit helped industry by generating mass demand, lowering unit costs, and raising productivity. But, he warned, increased consumption must be kept in line with increased exports and accumulated savings. Otherwise, trade imbalances would plague the resource-poor nation. As a cautionary tale, he cited press reports of a schoolteacher ruined by installment debt. His solution was to distinguish good installment buying of expensive durable goods from bad credit for food and drink.[100] The opposition between productive credit versus economically and morally suspect "consumptive" borrowing, for decades common in the United States, was taking root in Japan.

Others were decidedly more positive, noting that credit would promote social discipline along with economic growth. For such commentators, installment selling was "a great weapon for increasing demand and blazing the trail to new markets" that "clearly brings planning into daily life and regulates consumption."[101] In these views, spending and savings formed a connected bundle of virtuous "cultural" behavior: one sewing machine company called its layaway-installment plan "cultural savings."[102] The Tokyo Chamber of Commerce prefaced its survey of 1957 by describing the congruent interests of manufacturers, sellers, and consumers: "The installment sales system helps first of all manufacturers, but also retailers, to expand commodity markets. It plays an extremely significant role in sales strategies to insure stable sales volumes. In addition, of course it brings the benefit of raising the standard of living to consumers and rationalizes consumer outlays. It is not only the installment purchaser who enjoys the benefit of the practice. Even the cash purchaser benefits from the cost savings enabled by mass production."[103] Academic authorities such as Yajima Yasuo of Waseda University echoed this upbeat understanding that consumer demand boosted industry and reinforced social discipline. In a prolific body of work from the 1950s through the 1980s, Yajima argued consistently that installment selling was a key to expanding mass production as he described a virtuous circle of demand for consumer goods fulfilled through installment sales, in turn raising demand through a demonstration effect.[104]

All these voices in both institutional and academic reporting and in mainstream journalism were male. Since the 1930s, the overwhelming majority of users of some of the most popular goods sold on credit, especially the sewing machine, jewelry, and pianos, had been female.[105] As elsewhere, the history of installment selling in Japan was gendered. But the cultural construction of the female shopper differed from that found elsewhere, and it differed decidedly from the American discourse. As consumer credit in the United States reached giddying heights in the 1920s, anxious critics had depicted women as the source of the credit problem, vulnerable by nature to credit addiction and abuse.[106] But even in the most critical prewar assessments of installment credit in Japan, the scornful view of women as self-indulgent, undisciplined shoppers was not a major theme. Nor was it prominent in the postwar discourse. Rather, the marketing of the sewing machines on credit connected women with the twinned modern virtues of sober rationality, prudence, and investment on the one hand, and the pursuit of abundance and pleasure on the other.

In early postwar accounts, women appear as household managers with greater power to make decisions than in the past. The Janome Company's manual for salesmen in 1937 had urged canvassers to appeal to the interest of the husband and the entire family with the productive and educational value of the sewing machine. A 1963 panel of Janome salesmen claimed emphatically that the key to selling was to cultivate the housewife who, after all, held the power of the household purse.[107] But neither industry nor independent sources saw any need to worry about the fact that women were making the household spending decisions. They depicted women not as creatures liable to excessive spending out of ignorance about compound interest or prone to decadent hedonism out of the desire for useless baubles, but as wise and prudent consumers.

The problem, in fact, was that women too often found themselves victimized by irresponsible men. Counselor Satō presented women as superior to men in their credit-worthy self-discipline. It was the unmarried young men who were irresponsible and prone to go wild on credit spending.[108] Because ticket companies often sold their products through large companies to male employees, married salarymen could use ticket credit unbeknownst to their wives, especially if they purchased services rather than goods that they brought home. The wife only discovered these purchases when a pay envelope came home filled with cancelled tickets and no money, leading to "domestic struggles."[109] Because housewives were the most active members, Satō found that neighborhood associations were more reliable intermediaries than employers between ticket companies and individual consumers. Rather than fall behind—and suffer the scorn of their peers—these women would "go pawn something the night before, to be able to make the payment that day."[110] The *Asahi* in 1956 likewise noted that ticket companies could rely on female customers in neighborhood groups. "From the standpoint of collective responsibility, these women are really fierce gossips. If you fell behind, the neighbor group was unrelenting; one makes the payment on time, whatever it might take."[111] The *Asahi* in 1957 worried about the husbands, not the wives; the men tended to go irresponsibly overboard on the easy credit of tickets for food and drink. The wives could only lament at month's end that "his pay envelope is just full of ticket receipts."[112]

Most remarkably, one finds in the Japanese discussions of the 1950s not an American-style construct of the respectable woman as dangerous shopper, but the opposite: the stereotypical dangerous woman as a safe shopper. Counselor Satō ends his series of columns in 1958 with an anecdote presenting bar hostesses who cautioned their customers not to overspend their budgets as exemplars of "guidance toward a proper consumer life." A panel of Janome installment collectors in 1962 discussed how those who worked districts where many bar hostesses rented apartments had trouble finding these women at home and getting them to answer the door—they were always either asleep, in the bath, or putting on makeup.[113] But the tone of the discussion was jovial; no one suggested steering clear of such women, and nobody worried that they might default. The collectors simply advised each other to be disciplined, firm, and assertive at their jobs as they, in turn, disciplined this modern female consumer, a working woman who purchased her sewing machine to keep herself properly dressed on the job at low cost.[114]

. . .

Although it is tempting to assume that exports were the source of Japan's postwar "economic miracle," social scientists and historians have convincingly and for some time now been arguing that domestic demand was the key to postwar growth. Chalmers Johnson in his classic study, *MITI and the Japanese Miracle,* held that "Japan's growth did not depend nearly so much on exports as it did on the development of the domestic market."[115] Simon Partner has made the case that domestic demand for consumer electronic goods such as television drove the economy to expand in the 1950s and 1960s.[116] Most recently, Scott O'Bryan has shown that a Keynesian vision of the positive force of mass consumption framed economic thought over this span.[117]

In the broadest strokes, the story of the sewing machine as mechani-

FIGURE 31. This illustration, titled "Installment Serenade," appeared in the *Mainichi shinbun* on October 25, 1959, with a story on the spread of monthly installment credit for goods and services extending "from cradle to grave." It was drawn by Nasu Ryōsuke (1913–89), well known for his sharp and clever political cartoons. Nasu brilliantly depicts the mix of excitement and anxiety that accompanied the getting and spending of these years. (*Mainichi shinbun* and Nasu Miyo)

cal phoenix confirms these views. It also reveals a link between prewar and wartime developments and the postwar consumer revolution. The diffusion of the sewing machine continued from prewar through wartime to postwar years, first thanks to Singer and then to domestic producers. This good brought with it an American way of selling that served as a model for retail sales in other industries—electronics and home appliances in particular. The consumer of these goods, and her family, were encouraged and, for the most part, trusted to buy responsibly now and pay later on behalf of themselves and the larger cause of national growth and power. This was a Japanese variation on a global, if American-led, theme: a local version of the politics of productivity and middle-class consumption that carried a powerful ideology of the disciplined but happy consumer.[118]

Two illustrations that appeared virtually simultaneously in the United States and Japan give a sense of the mix of anxiety and exuberance at

FIGURE 32. On the cover of its issue for August 15, 1959, the *Saturday Evening Post* offered this romantic portrait of the postwar American dream by Constantin Alajalov, titled "Moonlit Future." The iconography is virtually identical to that of the *Mainichi* illustration, although the factories and the financing that made the dream possible are nowhere to be seen. (Curtis Publishing Company)

the heart of this Japanese story as it moved beyond the sewing machine and the radio to embrace a cornucopia of goods. The Japanese illustration (figure 31), drawn by Nasu Ryōsuke, a well-known political cartoonist, was published three months after the American one (figure 32), which appeared on the cover of the *Saturday Evening Post*. Perhaps Nasu, who worked primarily for the *Mainichi*, had seen the *Post* cover and was playing upon it. Just as likely, he was drawing on well-accepted images that constituted a globally circulating iconography of modern consumer life. The two artists position similar icons in quite different ways. A short description inside the *Post* does note of the couple that "to whatever extent they amass worldly goods, they'll share their happiness with the manufacturers and distributors," but the cover effaces the factory in its dreamy depiction of romantic consumer bliss.[119] The Japanese home is more crowded, and there is no swimming pool or car in sight, but the other goods are identical. And, most importantly, the *Mainichi* piece, titled "Monthly Installments at a Crossroads: Spreading from Cradle to Grave," explicitly completes the circle from consumer demand, through installment financing, back to the smoke-belching factory, and it clearly depicts consumer life as a source of both discipline and pleasure. A figure labeled "installment boss" stands gloating in the upper left corner. Nasu implies that this boss is skimming off an unfair share, and he renders the wage-earning father as stretched unhappily past his limit. But cheerful sounds appear to emanate from both the children at play and the wife at the stove. The consuming family remains the smiling engine for a rapidly growing industrial economy.

The sewing machine does not appear in either of these illustrations. In the United States it was no longer an object of high desire; it had been so widely owned for so many decades that its presence was taken for granted. Even in Japan, as ownership rates in cities by 1959 approached three out of four families, its place as the emblematic modern machine was less prominent than immediately after the war. But it is undeniable that the *mishin* was among the most valued objects sought after and used by women and families as they first literally and then figuratively rebuilt their homes and lives out of the ashes of war. For these women and their families, machine sewing was both a survival skill and a ticket to the dream of a bright life. In the 1950s and 1960s, sewing machines, sewing, and dress figured importantly in the emergence to cultural and social prominence of two emblematic types of women in Japan. One of these, the female consumer of a bright new life at home, was rendered by sellers, state officials, and journalists as the responsible shopper in a Japanese version of the demand-driven political economy of the postwar world. We now turn to examine her close relation, the "professional" housewife, a woman who played important productive and reproductive roles at the center of postwar Japan's middle-class society.

CHAPTER 7

A Nation of Dressmakers

As Japanese producers rebuilt their industry, the several hundred thousand women whose sewing machines had been destroyed during the war offered a ready initial source of demand, but the subsequent takeoff in sales was fueled by millions of new buyers. What enabled this surge appears inevitable in retrospect, but was not obviously in the offing at the time: the breaking of a demand-side bottleneck. Before and during the war, the *mishin* market had been constricted by the still modest spread of Western dress to adult Japanese women, which limited the interest in Western sewing and sewing machines.

The breakthrough took place from the late 1940s into the 1950s, but the pressure had been steadily, if gradually, building. Women were beginning to turn to Western dress in the 1920s and 1930s, and at first glance it is logical to argue that the war interrupted this shift: it led women to put on indigenous *monpe* outfits. Against this backdrop, the swift turn to women's Western dress in the early postwar era appears to mark a sudden, happy abandonment of wartime practice.[1] But a "leapfrog" story of a wartime halt in the modernization of dress followed by the postwar return to modernizing trends is too simple for several reasons. First, the turn to the *monpe* sustained modern values of efficiency and ease of movement; these pants were said by advocates to be both indigenously grounded *and* rational. Second, discussion of the wartime dress mode even included elements of modern concern for fashion. In both these ways, the wearing of pants for the war effort by city and town women was a potential bridge to Western dress.[2] Third, the wartime transformation of daily life offered the grounding for a postwar shift away from the kimono or *wafuku* and toward Western dress because the plebian *monpe* repudiated *wafuku*. The *monpe,* that is, sat outside the circle of what prewar reformers—and women themselves—understood to be *wafuku*. "Active wear" *monpe* triumphed over the Standard Dress of reformed kimono as garb for women in wartime. This greatly weakened any possible claims by postwar dress reformers that *wafuku* offered a platform for modernized yet Japanese-style clothing. Finally, the experience of war on the home front reinforced the importance of home sewing, and in postwar Japan as well, needlework at home remained a central and unusually time-consuming activity. It was one of the defining practices of the professional housewife at the center of the consumer economy and the social order.

LONG HOURS OF SEWING: THE TIME ALLOCATION OF THE POSTWAR HOUSEWIFE

The time surveys mentioned in the introduction found that urban women in Japan of the early 1950s typically devoted about three hours each day to home sewing.[3] As a comparative point of reference, surveys from the United States find that women averaged roughly one hour a day of needlework in the 1920s, and a good bit less in the postwar era. French women in the immediate postwar years appear to have stood only somewhat closer to their Japanese counterparts; according to a 1948 survey, they devoted one hour to ninety minutes daily, depending on family size, to "sewing, mending, knitting, and wardrobe management" (a more capacious definition than used in Japanese surveys). Data collected in 1937 by the famous British organization, Mass Observation, found that seventy middle- and working-class housewives sewed on average just twenty-eight minutes per day, and a 1949 study of "working-class wives in London," modeled on the French survey of 1948, tabulated merely twenty minutes per day given over to "sewing."[4] All of these surveys excluded home-based commercial sewing; other surveys, discussed below, make it clear that women in Japan did considerable sewing for the market, ranging from piecework for brokers to the sale of custom-made clothing directly to neighbors and the local community. There is no reason to think that home-based commercial sewing in Japan was *less* common than elsewhere, and it might well have been more widespread. In the aftermath of war, then, in both absolute and relative terms and for a variety of purposes, women all over the nation devoted long hours to sewing.

It is impossible to trace with precision the ensuing shifts in women's allocation of time in their daily routines, for in later years similar surveys asking comparable questions were not undertaken regularly. Although NHK reprised its ambitious wartime surveys of time use in 1960 and 1965, the compilers did not break out the category of "sewing" as they had in the early 1940s. But its survey of 1970, covering about three thousand women, once more asked about needlework. Hours of sewing had fallen sharply. Excluding any commercial homework, adult women, both those employed and those defined as "housewives," spent just thirty-three minutes on average each day sewing while devoting two hours and ten minutes to food shopping and cooking. Full-time housewives allocated nearly three hours daily to cooking and close to one hour to needlework, far below the stitching time of the 1950s. Although still high in comparative perspective, Japanese women's home sewing time declined significantly from the mid-1950s through the 1970s.

It is difficult to determine the pace at which sewing diminished in women's daily routines. It could have been slow but steady, or relatively sharp from a certain watershed moment. Most likely is that a slow decline in home sewing began in the mid-1950s, and a steeper drop-off came in the 1960s and 1970s as ready-to-wear dress became more widely available, affordable, and of better design. A 1959 study offers some suggestive evidence that sewing remained quite significant at this point. Surveyors of the Labor Ministry's Women and Children's Bureau interviewed 1,863 women nationwide, examining in particular "the attitude toward free time among housewives." The tricky aspect for our purposes is that the survey was not mainly concerned with housework and paid labor. It asked detailed questions only about the portion of women's waking hours that they considered "free time."

Women were asked only one question about housework: "How many hours a day do you devote to household work such as cooking or laundry?" As they replied, they had to decide for themselves whether sewing was a part of "household work" covered by the umbrella "such as." Certainly many did so, or would have been told to do so if they asked the interviewer to clarify, for a 1954 survey by this same bureau had explicitly noted sewing as an element of "housework."[5] The most common replies were "three to five" and "five to seven" hours of housework (respectively, 34 and 31 percent of all), and some of this time must have been spent sewing. The survey reported that 62 percent of these women owned sewing machines, while just one in four had washing machines, and only 16 percent had a television in the home. In addition to any needlework done under the rubric of "household work," and echoing the results of Japan's first time-use survey, conducted in Osaka in 1922, it is noteworthy that sewing ranked as the third most popular "free time" activity after reading and listening to the radio.[6] Asked how they primarily used their average of two hours of free time daily, about half of the women said they read, just over one-quarter listened to radio, and another one-quarter sewed. Playing with children was the primary activity of only 13 percent.[7]

The years of the takeoff of mass consumption of branded industrial goods in Japan thus coincided with the high tide of home sewing. Already a widespread practice in homes of the interwar era, household sewing had been boosted by the exigencies and policies of wartime; it flourished further and for a different balance of economic and social reasons after the war. Insofar as one understands consumption in modern societies as giving pleasure or fulfilling desires as well as satisfying basic material needs, the place of sewing in 1959 as a significant "free time" activity is important. Both for family consumption and for the market, both as "housework" and for pleasure, Japanese women in the 1950s and 1960s constituted a nation of dressmakers.

This activity was part of the daily routine of the "professional housewife" in her heyday. She was the Japanese incarnation of the full-time housewife who emerged to prominence in capitalist societies around the world in the mid-twentieth century.[8] This label is at first glance an oxymoron. Professionals are those who work for pay, but the labor of the female household manager is unpaid. I use it, following others (e.g., Suzanne Vogel), because it better renders the Japanese expression *sengyō shufu* than the usual translation of "full-time housewife."[9] The character for *gyō* conveys the sense of occupation or job, and the term *sengyō* means "specialized enterprise or occupation."[10] A *sengyō shufu* undertakes her work not only "full-time" but with a professional commitment to being a housewife on a par with the workplace professionalism of her salaried husband.

The concept and the social role of "housewife" (*shufu*) came into use and practice from the late nineteenth century through wartime. The term conveyed to a mass audience the sense of the good wife and wise mother constructed in the ideology of late Meiji times. In the interwar era it was given wide currency through magazines, most famously *The Housewife's Companion,* and this role held diverse meanings for those who prescribed it and those who lived it. The term *professional housewife* came into common use only after World War II. Through the decades of recovery and rapid economic growth, it came to be understood as the ideal role for a modern married woman.[11] Her vocational training and obligations extended from the management of home finances—and in some cases a stint or two supplementing family income as a homeworker—to housework, child rearing, and education. It included "care for the physical, emotional, and developmental needs" of the husband, children, and often elder parents.[12] Through tasks such as cooking and sewing, nurturing work was performed in deed as much as in word. By examining the ways in which the sewing machine became the everyday tool of Japan's every woman, we gain insight into the enduring life of the "professional" housewife.

WHAT WOMEN WORE:
THE END OF THE "TWO-LAYERED LIFE"

In 1957 one of Japan's most thoughtful and prominent social commentators, Ōya Sōichi, declared that a "dress revolution" had brought an end to the "two-layered life."[13] This decisive shift to Western dress unleashed a flood of demand for the sewing machine; by 1960 nearly three in four Japanese households owned one. The nation of dressmakers had been mechanized.

The change that Ōya noted did not start immediately with the war's end. For one year at least, city women continued to wear *monpe*. Famous photographs show them dressed in these trousers, crowding trains headed for the countryside, where they bartered for food, often trading precious kimono for something to eat. For some time their dress options were constrained by scarcity and rationing. The wartime rationing of the Japanese government was for several years replaced not by a free market in goods but by a new rationing system overseen by the occupation authorities. Through 1948 access to new fabric was strictly limited, and daily wear was a product of desperate improvisation; women patched or refitted old clothes, even converting military blankets to overcoats. One source of new dress was the stock of surplus wear of the occupation troops distributed to Japanese laborers and repatriates. Another was imported, second-hand clothing and charitable donations, especially from American church organizations. Only from 1949, with the recovery of domestic textile production, did significant supplies of fabric become available. Fabric rationing did not end until 1951.[14]

As home sewers and the mass of the population improvised their daily wear in the late 1940s, the prewar-to-wartime veterans in the movements of dress reform beat the familiar drums of "improvement" in one of several campaigns of women's organizations and state ministries aimed at "rationalizing" daily life. Under the rubric of the New Life Movement, these efforts initially focused on birth control to relieve population pressure on the struggling economy, hygiene in food preparation and kitchen design, and frugality in expenditures and in the use of scarce resources, including fabric.[15] The reform project's focus on dress began in earnest in 1949, when most of the major players in wartime projects for National and Standard Dress joined to found the journal *Ifuku bunka* (Clothing culture). Its manifesto reflected the stance of experts telling ordinary people how to improve their lives: everyday wear must draw on all sorts of research and thinking, that of "philosophers and historians, efficiency experts, scientists, economists and scholars of industry, as well as studies of aesthetics, custom and fashion."[16]

But before this advice spread in print, change in clothing practice was already underway. It took the form of a rush by women to put on Western dress, even in the face of scarce fabric and rationing. In 1955, Kon Wajirō's prewar colleague, Yoshida Kenkichi, now in his own right a pundit of dress, penned a compelling retrospective essay in a popular monthly: "From *monpe* to the 'one-piece' [dress]. [At first] the situation for clothing, food, shelter, was beyond impossible. I believe it was around the start of summer the year after the defeat when it became possible to survive, still with extreme difficulty, starting from food, to shelter and clothing. Because, as it started to get hot, women changed from *monpe* to skirts. . . . By 1948, it was clear that women's fashion [in Tokyo] had turned to the skirt, blouse, and some dabs of rouge."[17]

Kon himself wrote as follows of his summer 1950 explorations of the Japanese countryside: "I traveled through farm villages from Kansai to Tōhoku. What surprised me was that in whatever remote mountain valley, the women who gathered were almost all in Western dress. Of course the young women, but also the middle-aged and even the grandmothers, every one in Western dress. Among 100 people, there [were] perhaps one or two *kimono* in the mix." He commented that the outfits were simple dresses, not very stylish, but made from print fabrics new on the market. No longer was women's wear fashioned out of second-hand clothes. Even the agricultural cooperatives were selling these fabrics for Western wear—simple blouses, skirts, one-piece dresses. A few sewing schools had even appeared.[18]

Kon seems to have exaggerated the speed and totality of the shift. The summer of his farm visits, the *Yomiuri* conducted a nationwide poll and found that 61 percent of respondents still wore both Western and Japanese dress, while 29 percent had turned completely to Western wear. Despite *Yomiuri's* impatient interpretation that "old customs are not yet overcome," even these totals represent a sharp change from prewar or wartime surveys. The paper describes an energetic alliance of daily-life reformers, the state, and the media aimed at the renovation and rationalization of daily life. This alliance had survived the war and now sought "to give specific form to democratization."[19] But the account also suggests that changes were taking place in widely dispersed locales outside the control of coordinating bodies.

Indeed, the sources of new fashions in city, town, and country among women of all ages were various. They included patterns featured in the popular women's magazines as well as the outfits worn by the *pan-pan* (girlfriends of American GIs, or their prostitutes). In 1946 "stylebooks" featuring patterns and instructions on how to make Western wear began to sell in great numbers, presenting eager readers with the latest in American trends. They conveyed a complicated mix of desire and defensiveness, and a nonlinear course of change. The August 1947 founding issue of one of them, *Amerikan Fuasshon* (American fashion), acknowledged that "with new fabrics or accessories so difficult to obtain, we must resign ourselves for some time to the need to make do and improvise with whatever we have at hand." But, "whatever the materials, what we want to make are stylish Western clothes that fit our bodies perfectly." In meeting this desire, the fashion writers faced a tension between their understanding of Western styles and Japanese bodies. A certain skirt was said to "be a problem for Japanese to wear." A sun-dress slightly exposing the breast represented "a different concept from *wafuku.*" A certain chemise dress was "a look very difficult for Japanese to achieve, but what a free and easy way of dressing it is!" This magazine pledged to bring all of its content, the latest in summer and fall fashions "by airmail from the United States." Yet the editors frankly admitted to readers that they were selling not an obtainable reality, but dreams: "Wearing this sort of bold dress to a dinner or dance in today's Japan is unimaginable, but even the fantasy offers pleasure."[20]

Within a few years, large numbers of women gained the ability to act out their fashion fantasies. Introduced via the United States, Christian Dior's "New Look" long skirts of 1947 took off in Japan in 1948. From 1953, French designers, led by Dior, began to introduce their new spring lines in person in Tokyo, and over the following years, trends in women's wear in Japan were integrated into a global world of fashion. As this happened, the path from Paris to the Tokyo runway, and then to the wardrobes of millions of women, ran less through garment factories and stores with racks of ready-made dress than through the hands of home-based family sewers and dressmakers. This path is reflected in the results of a 1953 survey conducted by researchers at Chūō University that asked, "How do you learn of the latest fashions?" Almost nobody mentioned department stores. A clear plurality, 33 percent, replied "from stylebooks." These women were either making the clothes themselves or bringing the patterns to a dressmaker. Overall, in a literate nation of voracious readers, the role of print media was huge. Second to stylebooks stood fashion magazines, the source for 25 percent of the women, followed by newspapers at 15 percent.[21] Underlying and enabling this rapid change of clothes were the spread of new materials, especially nylon and other synthetic fabrics, and the spread of a machine that moved from object of desire to familiar tool for millions of women sewing at home, whether for themselves and their families or for neighbors and strangers.

With the easy wisdom of hindsight, this shift to Western dress seems inevitable. Yet I believe it requires explanation, for one can imagine alternative responses. The anxious tone in the fashion magazines promoting American styles in the late 1940s suggests that the editors feared women might be swayed toward *wafuku* by the difficult fit between bodies understood to be specifically "Japanese" and fashions that were both foreign and produced by a recent enemy. Perhaps they would return with delight to wear the familiar and comfortable *wafuku* they had reluctantly abandoned for the country bumpkin's *monpe*. Especially as the economy began to recover, a postwar search for elegance and pleasures denied in the worst years of war might have led back to kimono.

In fact, some women did follow this course. Yoshida Kenkichi's main theme in his ten-year retrospective was the move to Western dress, but he cautiously noted a countervailing leitmotif that he called "the reactionary revival of *kimono*," which began in 1948–49.[22] Yoshida did not elaborate on what happened in those years in particular, but as Liza Dalby notes in her cultural history of Japanese dress, the "*kimono* was resurrected in the 1950s and continued to grow in popularity during the 1960s," in what came to be called a "*kimono* boom."[23] References to the renewed popularity of Japanese dress appear in the popular press from the start of the 1950s. As the summer of 1950 approached, the *Yomiuri* reported such a fad for lightweight cotton *yukata* that department stores couldn't keep up with demand.[24] Showing a "new sensibility" marked by strong color contrasts and abstract design, kimono reportedly sold in department stores in 1955 at the rate of four for every six Western dresses.[25] In the summer of 1959, the lumber and furniture industry reported surging demand for paulownia chests for storing silk kimono. The call for these containers came from "city women in their thirties and forties gradually returning to *wafuku*."[26]

The postwar advocates of Japanese-style dress who promoted and celebrated these trends typically wrote in the genre of "theories of the Japanese" (*nihonjin-ron*), which claimed insight into a separate and superior cultural essence. They defended the unique suitability of *wafuku* to a unique Japanese culture. Higuchi Kiyoyuki, for instance, an ethnographer-historian cut from a far different cloth than Kon or Yoshida, began his career in wartime as an archaeologist of ancient Japan. He emerged in the postwar era as a leading spokesman for all manner of things traditionally Japanese. Writing in the 1970s, he saw the kimono as "functioning in complete harmony with the climate, social structure, physical body type, and aesthetic preferences of the Japanese." The sleeve opening under the shoulder was uniquely suited to ventilation in Japan's humid summers. The tight *obi* waist-wrap fought the particular tendency of Japanese women to develop pot-bellies! And the underlying aesthetic of the kimono conformed to a unique cultural preference for the "indirect expression of emotion."[27] Such reasoning in the end surrendered the functionalist ground of an argument that kimono or *wafuku* could be (re)made to fit the demands of modern life. It placed kimono in a museum-piece status as the high-cost carrier of traditional cultural identity, an item whose usefulness was not the main issue. The *wafuku* revival might have filled the paulownia chests of young women with kimono, but they needed to attend special classes to learn to put them on. In 1963 one sour commentator lamented this "ugly kimono boom." Young women had no idea how to carry themselves in clothes that they "wore about three times a year" to weddings or for New Year outings.[28]

In sum, as Nakayama Chiyo writes in her definitive account, despite periodic revivals and booms in Japanese dress, women "did not return to a life centered on *wafuku*" after World War II. They moved from *monpe* through second-hand American surplus goods to their own fabrications of Western dress. Thanks to *monpe,* they had begun to wear trousers *en masse* earlier than women in the West.[29] The reasons for the triumph of Western-style dress were surely several and related. Western dress was gaining adherents at a rapid pace already in the 1930s. During wartime, the dynamics of dress reform and the decisions of women to adopt *monpe* eroded the ground on which postwar advocates of a reformed Japanese dress might have stood. In addition, as Japan's mainly American occupiers proved themselves, for the most part, not to be the demonic figures of wartime propaganda, and as they brought attractive promises of democracy and peace, the appeal was profound of a wider range of cultural forms—dress included—that had long been associated in Japan with America and modern life. The legacy of a prewar American modernity in Japan combined with the improvisations of wartime modernity overwhelmed the anxious and nostalgic postwar advocates of Japanese-style alternatives.

The wartime and early postwar years put in place a feedback loop between home sewing, new fashion trends, and the consuming and producing household, whose implications we now explore. New modes of women's dress that made the sewing machine and its mastery indispensable reinforced the place of women as the managers of postwar consumer life. A boom in sewing education built on prewar and wartime foundations to spread a high level of skill at sewing and dressmaking to a remarkably high proportion of women. Often pursued and understood as a form of bridal training, these skills fostered the ascendance of the "professional" housewife in postwar society and culture.

TRAINING A NATION OF DRESSMAKERS

As women moved from *monpe* to the "one-piece" dress, sewing education became a big business. Some studied so they could fashion their own clothes from stylebooks. Those who purchased custom-made wear supported a burgeoning population of "Western dress shops" that relied on a growing number of dressmakers. In Tokyo, the number of such stores rose from thirteen hundred in 1943 to fifteen thousand in 1955, so that the city boasted the extraordinary density of one dress shop for every 120 households.[30] Ninety percent of the dressmakers who sewed for these stores were women, working on contract from their homes or in small workshops attached to some stores. In the 1950s, the commercial production and sale of women's wear relied on such dispersed networks of women sewing at home as much as it did on specialized centers of garment production.

The latter existed, to be sure. One of the largest grew out of the black market in front of the Gifu train station, about one hour east of Kyoto, employing seamstresses from the surrounding farming villages. By 1957, Gifu had seven hundred garment-producing firms, whose employees ran sixty thousand sewing machines. Prewar clusters of small-to-medium-scale garment producers reemerged in Tokyo, Osaka, and Nagoya, and they appeared for the first time in Niigata, Kanazawa, and Kyoto. Most of the nation's work clothes and the preponderance of children's and men's wear came to be assembled in such factories and sold ready-made by the end of the decade. But these districts produced relatively modest amounts of women's wear.[31]

This system of production contrasted sharply to that in the United States, much to the frustration of Itō Mohei. In 1929 he had founded a successful sewing school that, in the postwar era, became a leader in training dress designers for the garment industry. In a 1960 piece in *Women's Review,* entitled "On the Long Wait for Ready-Made," he lamented that whereas the United States ranked first, with 95 percent of women's wear purchased ready-made, followed by Europe at 90 percent, the total in Japan was merely 40 percent. A full 60 percent of women's clothing was made to order. How could it be, he asked, that in a poor country, the majority of women chose the more expensive option? His answer was that Japan possessed a new abundance of dressmaking capacity, one that yielded a balance of quality and cost in the production of ready-made versus custom-made dress that favored home-based commercial sewing.[32]

Women in the postwar years, that is, had come to constitute a nation of dressmakers, seamstresses, and home-sewing housewives. For this to happen, it was necessary for millions of them to learn new skills. Although women in prewar and wartime Japan sewed for long hours, only a minority practiced Western sewing, and only some of them had access to sewing machines. The home economics classrooms of the public school system did not offer sufficient hours or depth of instruction to impart more than a basic familiarity with machine sewing, so the primary providers of this expertise were dressmaking schools. The leading institutions had been founded in the 1920s, and, alongside newly founded competitors, they now expanded at a pace that astonished observers. An extraordinary number and proportion of Japanese women learned to sew in these schools, whether to learn a proper lady's hobby or to acquire skills useful for a future career as household manager, seamstress, dressmaker, or sewing teacher in her own right. In 1958 the phenomenon inspired Ōya Sōichi to make the exuberant claim that "no foreigner who observes Japan's Western sewing boom is unsurprised," for "there is no place in the world where schools for Western sewing are as popular as in Japan."[33]

This training was part of the nurturing of the full-time, professional housewife of postwar Japan.

This schooling boom was built figuratively and in some cases literally upon prewar foundations and wartime ruins. In the last days of the war, sewing schools had been forced to close or cut back to minimal operation in temporary quarters. With just forty students remaining, the Cultural Dress Academy (Bunka fukusō gakuin, abbreviated Bunka) by the summer of 1945 ran classes just once a week in a barrack-like structure in Tokyo, but kept seventeen teachers on its staff. Ironically, after surviving the war, the school was forced to close for lack of a usable building in the fall of 1945. It regrouped one year later, recalling its teachers and recruiting a full house of three thousand students within several days of announcing it was re-opening. Running classes in morning, afternoon, and evening shifts to make the best use of its cramped facilities, the school saw its enrollment double to six thousand in 1947. It reached ten thousand students in a brand new Tokyo campus by 1955, at a time when the largest sewing school in the United States was said to enroll no more than five hundred students.[34]

Bunka and its successful competitors, notably Sugino Masako's Dressmaker Women's Academy (Doresu meekaa jogakuin, abbreviated Doreme), grew to even greater size and profitability by opening franchised schools run by their graduates throughout the country (figures 33 and 34). Only loosely connected to the parent school, the franchises were expected to earn their own way. The franchise principals were invited to meetings to exchange ideas on best practice, encouraged to tune in to shortwave radio broadcasts of the parent school's classes, and expected to purchase textbooks and other teaching materials from the prolific and profitable Bunka (or Doreme) publishing division. By 1958, Bunka recognized 300 such schools, while Doreme claimed 700.[35]

Already in 1947 400 sewing schools nationwide enrolled 45,000 students, and these totals soared to 2,400 schools and 360,000 students in 1951, and as many as 7,000 schools with 500,000 enrolled by the mid-to-late 1950s. Roughly 900,000 girls were graduating from middle school in each of these years; about one-third of them went on to high school. It is clear that attending sewing schools for a time upon graduation from either middle or high school was a postgraduate course that virtually all young women must have thought about and as many as one-half actually followed.[36]

The students who graduated with relatively strong skills brought to their home sewing a quasi-professional commitment, whether they fab-

ドレメ通り　午前九時の始業を前に道路をうずめて流れる女性群。一日五千名が集まる

FIGURE 33. This photograph of the rush hour of students heading for their classes at the Doreme sewing school vividly conveys a sense of the mass character of postwar sewing classes. It was published to illustrate Ōya Sōichi's essay on the 1950s boom in sewing schools in *Shūkan Asahi* January 26, 1958. (*Asahi shinbun* Photo Archive)

ricated for themselves, their family, or the market. Ōya brilliantly rendered this serious and pragmatic spirit in a 1957 *Asahi* column where he discussed the "goals of entrants to Western sewing schools." The energy of the entering students "is focused not only on the desire to wear fine clothes and look beautiful, nor is it a mindset of faddish imitation of others." The many young women who saw their studies as a "pre-wed" course in bridal training had a sober self-understanding. Some saw the acquisition of sewing skills and graduation from one of these schools as a way to move up in the marriage market "from the prospective bride of a low-ranked clerk in a provincial government office to the spouse of a supervisor or section chief." Others calculated that their tuition would

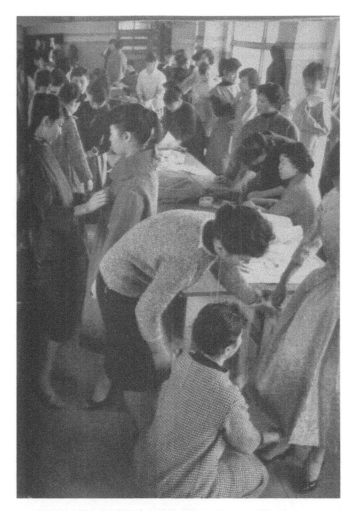

FIGURE 34. This second photograph from the spread that accompanied
Ōya's 1958 essay shows a classroom scene at Doreme, notable for its
tone of serious yet convivial endeavor. (*Asahi shinbun* Photo Archive)

be offset by the ability to fabricate the clothes for a trousseau in the classroom. Still others felt that no self-respecting future wife should let her
husband buy his suits off the rack: "I want to make him a suit myself."
Many others saw their studies in the prewar spirit of "self-sufficiency"
as a step toward economic independence, aiming to work as dressmakers or open their own shops. Ōya concluded with a flourish that a spirit
ranging from realism, practicality, or independence to feminine affection
and innocent dreams "cannot be pigeon-holed as either hardboiled or

sentimental. The sound of the sewing machines pedaled by the daughters of Japan is dynamic and complex."[37]

Prewar and wartime sewing schools had enrolled students aiming to gain marketable skills as well as those looking for a bridal "finishing" course or practical household management skills. The occupational track had dominated, and, given the relatively modest skills demanded by the industry at the time, one could learn enough to be employable after a short course of study. During the postwar years, standards for Western dress and the corresponding skill requirements rose. Without further training, the many women who undertook only the short basic courses at these schools were not prepared for careers as professional dressmakers or designers; for these women, sewing schools served more as bridal than vocational training, especially as skill in Western rather than Japanese sewing came to be seen in conventional morality as the incubator of feminine virtue.[38] But even these women had built a foundation for the acquisition of further skills if needed or desired. With a bit more practice or study, they could fabricate clothes for themselves and their children, demonstrating love of family while saving money. And the significant minority who pursued longer-term, more advanced courses of study indeed graduated with the ability to sew professionally, whether from the home, in a dress shop, or as a factory-based seamstress.[39]

This education of masses of women skilled in Western machine sewing and eager to put these skills to use at home, whether to make clothes for their families or for the market, thus furnishes a key part of the answer to Itō Mohei's frustrated question: Why did the production of custom-made clothing continue to dominate the domestic market for women's wear as the 1960s began? He began to answer his question by recognizing that the cost differential in Japan was relatively small. In the United States, he wrote, even a cheap outfit made to order cost nearly four times more than a dress off the rack. In Japan, it was only 30 percent more costly. For Itō, whose school hoped to place graduates in jobs as designers of ready-made wear, the low markup resulted from self-exploitation by Western dressmakers who "simply cannot afford to lose to ready made clothing, so they make unreasonable efforts to sell cheaply. They live in poverty, and their products are too cheap. If they don't sell cheap, they can't get orders. This is after all a country with excess population and too many competitors."[40] There is a kernel of truth to his repetitive complaint, but it does not sufficiently consider two aspects of the production system. On the side of industrial producers, ready-made dress had not yet shaken its reputation for poor quality; those wearing clothes "off the rack" were dismissed as *tsurushinbo,* a derogatory term with connotations similar to a racial slur.[41] On the side of the producers of custom-made clothing, the realm of dressmaking and the ownership of dress shops rested in the hands of women based in the home who were highly skilled but viewed their labor as supplemental to a family income. By combining skilled productive labor with the work of home management they were led, or forced, to work so "cheaply" that they could outdo the producers of ready-made clothing on quality for a modest premium in cost.

SEWING FOR THE MARKET

These dressmakers and seamstresses were among the many women who contributed to a counterintuitive social trend. In the immediate postwar years, it is no surprise that thousands of war widows and other home-bound war survivors desperately sought to make ends meet through home-work. Less expected is that as the Japanese economy began its renowned, high-speed growth of the 1950s and 1960s, all sorts of home-based production expanded significantly, with needlework in the forefront, even as increased numbers of homeworkers understood themselves as members of the middle class.

The Labor Ministry's Bureau of Women and Children (Rōdō shō, fu-jin shōnen kyoku; also the source of several important time-use surveys) took the lead in documenting this trend. The bureau was founded in 1947, to considerable fanfare and controversy.[42] In the summer of 1950 the bureau undertook its first study of homework, a relatively modest survey of fourteen hundred practitioners in Tokyo and three other regions. In 1954 the bureau organized two considerably more ambitious efforts, surveys in Tokyo and Osaka that drew on hundreds of volunteers (six hundred in Tokyo alone) from several women's organizations to assist the bureau's professional staff in sampling a full 1 percent of all households in each city. In 1968 the bureau carried out an even more ambitious survey of thirty-seven (out of forty-six) prefectures nationwide, including Tokyo and Osaka. These surveys usefully defined homework in expansive terms to include more than those who received materials and sometimes tools from a manufacturer, a broker, or a vocational center and then processed or manufactured a good at home, receiving a per-unit fee from the provider of the materials: in needlework terms, seamstresses. The bureau's definition also encompassed those who—so long as they did not maintain a storefront—themselves purchased the necessary materials, fabricated the good at home, and sold it directly to a consumer:

in needlework terms, dressmakers.[43] Comparison of these three major surveys, two at the start of Japan's era of high growth, and one near the peak of two decades of the "economic miracle," offers an excellent view of homework and the homeworkers.

In the mid-1950s, surveys show that in 6–10 percent of households in major Japanese cities, at least one member engaged in homework at any one time. In Tokyo in 1954, about one hundred thousand households, 9 percent of all in the city, included a homeworker.[44] The more broadly based survey by the Women and Children's Bureau in 1968 found that in 12 percent of all households surveyed, at least one person did homework, while someone in another 12 percent of households was hoping to do it. In 9 percent of households, at least one person not at that moment engaged in homework had undertaken it at some point in the previous five years.[45] Put differently, in roughly one in five Japanese households at least one person had been a homeworker for some portion of the span from 1964 through 1968.

These high-resolution social snapshots suggest that the extent of homework remained steady or grew slightly over the course of the 1950s and 1960s. Additional less granular nationwide data support this interpretation. Throughout Japan, homeworkers supplementing family income numbered roughly 700,000 through most of the 1960s. This jumped dramatically at the end of the decade, to nearly 1.3 million in 1969 and 1.6 million in 1970, reaching a peak in 1973 of 1.63 million homeworkers. Only in the following years did their ranks gradually decline.[46] A noticeable minority of the 1950s homeworkers were men: 15 percent in Tokyo in 1954 and 17 percent in Osaka in 1955. These men were generally ill, disabled, or unemployed. One can assume they were often war veterans. By the time of the 1968 survey, such men, if they survived, had apparently found work outside the home or retired; in that year only 1.8 percent of homeworkers were men.[47] This means that the overall increase in homeworkers reflected an even greater increase in the practice among women.

Needlework was by far the most common form of home-based production throughout these two decades. In Tokyo in 1954, "Western sewing" accounted for 16.5 percent and "Japanese sewing" for 13.2 percent of homeworkers. These workers, plus those doing piecework for shoes, underwear, hats, gloves, knitting, embroidery and other miscellaneous tasks using sewing machines, accounted for 47 percent of all homeworkers. One-fifth of all Tokyo homeworkers sold directly to the consumer, and the proportion of these independent homeworkers was especially high in the needle trades; fully half of "Japanese sewers" and close to half of those doing "Western sewing" or "repair of Western undergarments" sold directly to the customer.[48] By 1968 the proportion of needlework had increased; 61 percent of the homeworkers surveyed were engaged in "processing of textiles."[49]

The women in all these surveys were predominantly wives and mothers who combined homework with other responsibilities of childcare and household management. Most homeworkers from the early 1950s through the late 1960s spent 4–6 hours per day at their jobs about five days per week, considerably fewer hours than typical employees outside the home, whether in offices or factories. And, in contrast to the findings of interwar surveys, a significant minority of these homeworkers consisted of women in middle-class families. In 1954 Tokyo, a full one-third (34 percent) of homeworker families were headed by a husband whose occupation was either "company employee" (21.4 percent), "public employee" (6.7 percent), or "other salaried person" (5.9 percent). Only 12 percent of the male household heads were factory workers, and just 9 percent were self-employed. Fourteen percent of household heads were women, either the homeworker herself or her mother. The Osaka profile was a bit less middle-class; even so, one-quarter (24 percent) of households were headed by one or another of the "salaried" categories. The 1968 survey unfortunately aggregated the data in different fashion, but the middle-class proportion appears to have been comparable or increased.[50]

The homeworking woman's income was almost always supplemental to the family budget, but her proportional contribution rose slightly over the high-growth era; homeworkers in 1954 Tokyo contributed on average about 11 percent to a family income, while those in Osaka in 1955 contributed 14 percent. A 1966 survey of homeworkers in needle trades alone found that they contributed an average of 17 percent.[51] A related sign of increasingly middle-class status for these homeworking households was that their income level increased over time relative to the general population. In 1954 Tokyo, homeworker family income of ¥16,000 was just over half the average household's monthly income of ¥31,000. By 1975, a survey by the Ministry of Labor showed the income of average homeworker households had reached 80 percent of that of the average "working household."[52]

In sum, during the two postwar decades of rapid economic growth, the overall number of homeworkers increased, and these women provided an important supplement to family incomes. Homework was never the monopoly of the poor. Already in the interwar era, a modest group of aspiring middle-class families, represented by the tellers of "true tales" in women's magazines, was engaged in homework alongside a larger proportion of the working class. During the postwar era of high growth, the relative proportions of lower-middle and middle-class homeworkers increased. Their jobs of choice involved needlework and sewing machines.

The war cast its shadow in one way or another over every homeworker's story of the 1940s and 1950s. Yoshimoto Yōko ends her analysis of the "true stories" from *The Housewife's Companion* with a sidebar "true story," which she uncovered herself, of Matsui Yoshiko, the woman who invented an erasable paper that could be used with tailor's chalk in sewing design and pattern work. At the age of thirty-three, Matsui survived the horrific night of the March 10, 1945, air raid in Tokyo. The next morning, she returned to her burnt-out home to find that the sewing machine was the only object that survived the conflagration. When she later decided to divorce her husband and move back to her natal home in Kobe, she took the machine head with her, as the "symbol of her independence."[53]

The Japan Association of War-Bereaved Families (Nihon Izokukai), more famous for its political role in controversies over the appropriate way to mourn the war dead, also collected survivor accounts to bolster its call for state support for its members.[54] As reported in a hefty volume published in 1963, some women—and surely this was typical—began but then abandoned homework, finding it insufficient to make ends meet. Mori Kaneko's soldier husband died of illness in China in 1944. Until the war ended she supported herself by weaving at home. After the war, she learned to operate a sewing machine, but took in homework inconsistently, as her brokers could not provide steady work. Each morning she faced the sewing machine and offered a "prayer" of appreciation and "hope that I won't injure myself today." In 1953 she shifted to selling life insurance, another occupation favored by war widows, who could trade on sympathy and testify convincingly to the difficulty faced by someone after the death of an uninsured husband. She ended her memoir noting that this, too, was no easy job; collecting the premiums was especially tough.[55]

Matsuda Kazuko's struggle began when she was seven, with her father's death in Sumatra in January 1945. In 1950 her mother used a government loan available to survivors of the war dead to purchase a sewing machine and cutting table, and rented a room to start dressmaking. She learned to tailor men's clothes, but her earnings could not support the family. It was hard to keep up with changing styles and customer taste, and it was also difficult to keep up with housework. In 1953, Kazuko, now fifteen, began to study Western sewing at a sewing school. She helped her mother's business when she could, but they relied on welfare payments to survive. They managed to build a small home of their own, thanks mainly to Matsuda's survivor pension.[56] Sotozaki Yae's "twenty years of difficulties" ended somewhat better. Her husband died in 1939 in China. During the war, she worked at home, sewing Japanese-style clothes, lonely and desperate, crying all the time. A friend suggested she try to support herself with Western sewing, and the local authorities arranged for her to attend sewing school. She later apprenticed to a local clothing store, given jobs by the sympathetic owner. But eventually she opened her own shop and prospered.[57]

The truth, as told numerous times in *The Housewife's Companion* in the 1950s, read more like Sotozaki's account than the darker survivor tales. T. S. from Kyūshū, just returned from overseas, with a husband working in a hospital, bought a sewing machine with a lump sum of compensation for luggage lost en route home. To satisfy her customers, she needed to keep up with fashion trends in women's magazines and department stores, and she took care to undercut the custom-made prices of the latter by about 30 to 40 percent. In the end, she happily supplemented her family income to the tune of ¥2,000–3,000 a month by sewing part time while doing housework.[58] On the more entrepreneurial end of the spectrum in *The Housewife's Companion* stood a widow from Sendai identified as T. O. She and her husband returned home from many years overseas in 1944, but he soon died. She was relatively fortunate to have savings that she estimated could support her and her four children for three to four years, but with the long term in view, she enrolled in a sewing school. In 1948, her savings nearly exhausted, she opened a "research institute" (*kenkyūjo*) at her home. This term seems to have designated a combination school and dressmaking operation, probably labeled this way to escape taxation as a business or avoid the need to register as a school. By 1953 her home-based enterprise, with three sewing machines purchased on credit, boasted more than a dozen students coming for occasional lessons. With more orders than she could handle, her monthly income was ¥12,000, placing her comfortably on the lower rungs of the middle class, happily able to support her children's schooling.[59]

Such stories were a staple of virtually all the women's magazines of the time. In addition to occasional "true tales," *Women's Club*, the most popular competitor to *The Housewife's Companion*, ran advice columns on "Homework in the Major Cities," or "How to Seek and Choose Homework," as well as numerous ads for sewing schools and sewing machines. These of course noted the opportunity to earn through homework as an important reason to enroll or buy a machine. The sewing machine sellers filled their in-house magazines for the sales force, as well as glossier publications aimed at customers, with stories such as that of Kawaguchi Kinuko, introduced in chapter 6, who "managed to save the sum of one million yen simply doing Western sewing homework" with her reliable Janome machine.[60] And in the 1986–87 edition of its long-running series of morning dramas, Japan's national broadcaster (NHK) offered a semifictional version of the prototypical "true tale" of postwar dressmaker success, titled "Miyako no Kaze" (Breeze of the old capital). Whatever the medium, these stories constitute the public representation of the mainly private practice of home-based needlework; the authors, editors, or producers had good reason to spin stories of eventual success in order to sell magazines or win viewers. The mass of homeworkers often labored in difficult circumstances, paying a price for their earnings with their health or their time for family or other pursuits. The willingness of so many women, dispersed in their homes and also engaged in domestic labor, to work long hours for modest profits created the supply of custom-made dress of quality and cost sufficient to outsell the producers of ready-made women's wear, which so frustrated Itō Mohei.

The darker stories compiled by the Association of the War-Bereaved, while biased toward winning sympathy for their members' suffering, may well be closer to the majority experience of homework than the magazine tales. But to follow Itō in casting these women as an immiserated underclass misreads their story. They included both survivors and upwardly mobile strivers. The era of high economic growth coincided with a sharp rise in the numbers of homeworkers, and I believe that causal links ran in two directions. Certainly Japan's "economic miracle" was in part fueled by the low-paid labor of these women. But they acted on the demand side as well. In touch with a world of fashion and glamour, they undertook this work both to fulfill their customers' dreams of a bright new consumer life and to secure this life for themselves and their children. As it was used for homework, the sewing machine was not a tool that primarily immiserated its users, as Marx had it; nor was it a means to live a simple life, as Gandhi claimed. As it connected women and families to the consuming world, it was a tool of class integration more than division; it helped build the cultural hegemony of Japan's middle-class life from prewar through postwar years. And as women moved in and out of the roles of homeworker-housewife and full-time housewife, needlework helped link their identities as income earners and professional home managers.

SEWING FOR THE HOME

Home-based commercial sewing accounted for a significant portion of the household use of sewing machines in the early postwar decades. But the majority of *mishin* buyers and users stitched most of the time for themselves and their families. The time surveys of the 1950s, which found that women spent two to three hours at sewing each day, took pains to note that commercial sewing was tabulated in a separate, much smaller category. Across all households, the commitment of women to commercial sewing was a modest fraction of the time spent on family sewing.

Reflecting and reinforcing this allocation of daily time was a shift over historical time in the reasons most women enrolled in sewing school. As dress historian Nakayama has argued, in the postwar years sewing schools "changed from occupational to bridal training centers." Acquisition of the skills needed for a future as a housewife and home manager came to outweigh vocational training as the reason to enroll.[61] The Yokohama Western Sewing Vocational School, founded in that historic port city in 1927, was renamed after the war for its founder, Iwasaki Haruko, as the Iwasaki Academy (Iwasaki Gakuen). It grew to become the city's most important sewing school. Ohtsuki Toshiko and Takayama Kazuko were two of Iwasaki's veteran teachers, handpicked from among the most talented young women enrolled at the school in 1940 and 1948, respectively. Interviewed decades later, these women recalled that their students of the late 1940s had enrolled to learn a skill to earn money for themselves or their families. Takayama noted that many were widows with children, and they were desperate. Ohtsuki added, with a nostalgic note in her voice, that "the students were truly committed and serious. They had a spirit, a resolve born of desperation that was not there before or after. I can never forget those students." They would start doing homework while still students, taking home their class projects and selling them, remaking and reselling old clothes, and passing on their lessons to others at home.[62]

Such students remained common in the 1950s in their recollection, but in the 1960s and 1970s they observed that students on the "pre-wed" track became the mainstream. The latter opted for a six-month "short course" that allowed them to cycle through similar courses in cooking and perhaps flower arranging or tea ceremony as they or their parents looked for a mate.[63] This shift in the relative balance of commercial and family home-sewing appears to have begun early in the 1950s. Even in December 1951, a short column in the *Asahi*, "Shopping Notebook," ran a piece on the *mishin* that began by noting that "the huge jump over the past year in the sewing machine diffusion rate, from 16.7 to 30 machines per 1,000 people, is due to the well-known fever for Western sewing and the move of this good into the category of necessity for a trousseau."[64]

Monthly magazines and the daily papers energetically cultivated this growing audience of bridal-track graduates. Testimonial ads for sewing schools and sewing machines were not above playing on their fear of failure. One school ad that ran repeatedly in *Women's Club* in the late 1940s told the cautionary tale of a young bride who "wept because she could not do Western sewing."[65] In addition to monthly magazines for women, the women's page in daily papers devoted extraordinary space to family sewing columns. Alongside prose instructions, these typically presented detailed patterns for dresses, blouses, and skirts for women, along with all manner of outfits for boys and girls. Such columns far outnumbered the true tales of women who sewed for independence and income. *Yomiuri* in late 1952 inaugurated a five-year run of "Western Sewing Lessons" placed prominently on the paper's women's page every three days, 581 lessons in all. A sidebar to the first, "On the Opening of our Western Sewing Classes," pledged to offer basic knowledge and the latest fashions for the "edification of ordinary women with interest in Western sewing," drawing on the advice of "experts, without bias toward any one style" (that is, the styles taught by Bunka, Doreme, or any other school).[66] This feature ended without comment at the end of June 1957, replaced by a more elaborate but less frequent "This Month's Western Sewing," which ran sixty-seven half-page features from January 1960 through August 1965. The *Asahi* was a much later entrant, but in April 1959 it opened a "Wednesday Western Dress Shop" on the women's page, and through 1970 it ran 529 weekly columns of the same type as those in *Yomiuri*.

The frequency, prominence, and complexity of such columns in the daily press were distinctive features of home sewing in Japan. American women's magazines, especially in the early twentieth century, when home sewing was most popular, devoted considerable attention to sewing advice, as did Japanese magazines. Some daily American papers in the 1920s and 1930s ran syndicated features such as Clotilde's Practical and Fancy Needlework, but Clotilde's columns typically offered prose descriptions of fashion and sewing projects, only occasionally adding a simple pattern drawing. One finds nothing in the American press at all comparable to the outpouring of newspaper sewing instruction in Japan (see figure 35).[67]

This press support for home sewing reflected and reinforced the emerging dominance of Western sewing over Japanese *wasai* as both practical skill and preeminent feminine virtue. The 1959 survey of the Women and Children's Bureau made it clear that for many (one in four housewives in their study), sewing was not so much work as a "leisure" activity; put differently, one might say that many Japanese women approached their leisure in a remarkably practical or professional spirit. Three years earlier, in 1956, a survey undertaken by what Janome called "an authoritative research institute in Tokyo" offers nice insight into the self-understanding of home sewers. The survey looked at what it called "the psychology of buying," and aimed to answer the question, "What does a lady want to buy?" with particular focus on the relative attraction of various "household cultural goods."[68]

The respondents were asked to ponder the following story: "Akiko desperately wants to buy household cultural goods. She has saved ¥30,000 for this. She goes with this money to a department store and looks around at various goods. An electric washing machine is ¥25,000. A sewing machine is ¥26,000. A TV is ¥80,000, and a mixer is ¥12,000. What does Akiko most want to buy, and why?" The goal of framing the survey in this fashion, it would seem, was to posit a person who owned none of these goods in order to induce the respondent to state her own preference among items she already owned as well as those she did not yet possess. Just over half of the women surveyed (51 percent) believed that Akiko's greatest desire would be to buy a sewing machine, followed by a washing machine at 41 percent, and an electric mixer at 4 percent. Only 2 percent of the sample made the television Akiko's first choice.

The survey interestingly divided respondents into groups who actually owned particular clusters of goods. Of those who already possessed both sewing and washing machines (19 percent of all), 49 percent had Akiko buy a sewing machine, and 37 percent had her buy a washing machine, but of those who did not yet own either of these goods, 69 percent had her buy a sewing machine, while just 22 percent had her buy a washing machine. Janome's bulletin writers did not seek to explain this difference, simply concluding that the data overall showed that women saw the sewing machine as the highest priority daily-life necessity, so Janome sellers could go forth and knock on doors with confidence. But it seems reasonable to see this gap as a measure of the power of marketing and peer pressure to generate desire for this good. The lower priority given the *mishin* among those who already owned it suggests that a number of buyers were disillusioned once they actually owned a sewing machine; it is likely that they ended up not using it as much as they expected, and with greater difficulty than promised, so they had Akiko opt for a washing machine. Even so, this is only a relative diminution of desire; the preference for the sewing machine over all the other "cultural goods" is indeed impressive.

The respondents were also asked why Akiko made her particular choice. Among those who had her buy the sewing machine, the vast majority (91 percent) gave "usefulness, necessity" as the most important reason for buying the good, followed by "preparation for the future (6 percent), with the remaining handful noting "just because she wants it" or "no particular reason." The investigators probed further into what women meant by "usefulness" and "necessity." The most frequent elaborations were "it has a greater use value than other household goods" (18 percent), "because I can do Western sewing and homework myself" (17 percent), and "it is the most needed item for a woman in family life (10 percent). The first and last of these comments represented a plurality who saw the *mishin* as useful primarily in its contribution to family and home life (28 percent), while the second defined "useful" as a combination of value directly for the family and indirectly through the wages of the homeworker.[69]

Also in 1956, more than five hundred women sent letters describing the joys of sewing in response to an ad in *Asahi* placed by twelve makers of home appliances announcing an essay contest on the topic, "Which appliance is the best value?" A sample of prize-winning entries was published in Janome's in-house bulletin. Typical was Yabuta Haru, who wrote, "Basically, I just like to sew things like children's clothes, using magazine patterns." For Yabuta, purchase of a sewing machine saved time, was economically beneficial, and "gave us the extra margin that helped elevate [the cultural level of] our daily lives."[70]

The sewing machine offered women like Yabuta, as well as those who spoke for Akiko and those who in 1959 saw machine sewing as a leisure activity, a powerful combination of pleasure and practicality. A woman saved money by sewing for the family rather than buying custom-made or ready-made clothes. She felt pride and satisfaction in the pleasure taken by children or a husband in their dress, and she could make herself attractive as well. The home-based sewers of early postwar Japan gained satisfaction through their mastery of a skill understood to define the feminine virtues of nurturance and practicality and to validate a woman as a competent, responsible household manager.

That such women throughout Japan possessed this high level of skill as one element of their professional competence helps explain an intriguing puzzle in their behavior: their reluctance to use pattern papers. The American publishers of sewing patterns, Simplicity and McCall, entered the Japanese market with high hopes in the early 1960s—at first glance with good reasons. As Ōya Sōichi had noted in 1957, Japan boasted a reputation as a "world leader" in both sewing machines per household and, by an even greater margin, in the number of students enrolled in sewing schools.[71] What could be more promising for sellers of pattern papers that promised easy fabrication of the latest fashions for women and children? Yet, in spite of this seemingly huge pool of ready customers, Simplicity and McCall's fared poorly. By the early 1970s, at a point when six hundred thousand women in Japan were graduating annually from Western sewing schools, the two companies were selling only several million patterns a year, compared to four hundred million in the United States.[72]

The logic and hopes of Simplicity and McCall had failed to take into account the skills and professionalism of the sewing housewife or her young adult daughter. As Ohtsuki Toshiko and Takayama Kazuko of the Iwasaki Sewing Academy explained, their students saw these patterns as a short-cut beneath the dignity fostered by their training and skills. It was not that the graduates of sewing schools never used paper patterns. They simply did not need to buy them off-the-shelf from Simplicity. Projecting from small-scale drawings, with arithmetic skills and the ability to take their own measurements or those of a child or a customer, they were able to make their own paper patterns. Only the widespread possession of this ability can explain why mass-circulation dailies would devote precious space several days each week to the sort of small-scale drawing shown in figure 35. Indeed, teaching the skills to do this was at the heart of the sewing school curriculum. Ohtsuki and Takayama stressed the professional stance of the home sewer. They believed it was a pleasure and source of pride for the students to be able to work from their own materials and ideas to create clothing for themselves or their families, or as dressmakers.[73] The *mishin* thus blurred or crossed the class line. Its multitude of users moved between occupational and family sewing, and came to include huge numbers of middle-class sewers. The long-standing centrality of home sewing bolstered and was bolstered by the ascendance of the professional housewife.

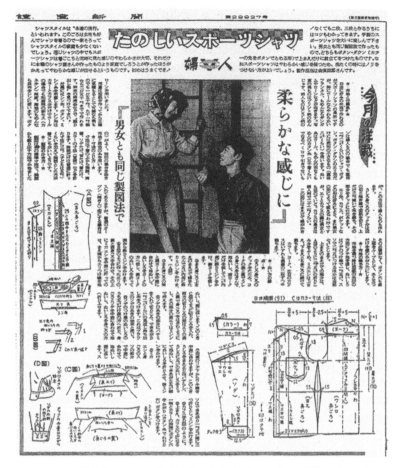

FIGURE 35. This installment in the *Yomiuri shinbun* column "This Month's Western Sewing" appeared on February 27, 1960, under the headline "Enjoyable Sport Shirts." In its use of detailed scale drawings, which readers would project for marking and cutting of fabric, it is a typical example of similar columns running since the early 1950s that presumed high skill levels among their numerous readers. It also offers a relatively new twist with a photograph suggesting that part of the enjoyment would lie in attracting the admiration of a young man. (*Yomiuri shinbun*)

THE END OF THE TRANSWAR ERA AT HOME

The first regular sewing column in the *Yomiuri* ran in 1952. The paper's last regular column, a weekly feature called "Yomiuri Western Sewing," which began in January 1967, was published in August 1975. *Asahi* closed down its "Wednesday Western Dress Shop" in 1970 with this note: "It's been eleven years since we opened for business on April 1, 1959, with the motto 'clothes that go well with everyone and that everyone can easily sew.' . . . Now in the era of ready-to-wear dress, we see from our readers' letters that the figure of the Western-style home sewer has gradually changed. The variety of materials has increased amazingly. An overflow of fashion information pours forth each day. We hope to see you again with a renovated dress shop appropriate to our age."[74] In the following years, *Asahi* ran only an occasional seasonal feature on sewing; no sustained reunion of readers and writers took place.

This lifespan of the newspaper sewing column—roughly a quarter of a century—marks the chronological bounds of Japan as a nation of dressmakers. It also marks the time when the professional housewife reigned most securely as the ideal for women. Her position as the exemplar of adult womanhood was never unchallenged; periodic "housewife debates" erupted in public life across these years, and we can be sure that many private conversations offered diverse answers to the question of how women should best lead their lives.[75] Even so, an enduring social and cultural formation runs from the 1920s across the war and through the high-growth era, marked in terms of women's roles by the prominence first of the prewar-to-wartime "housewife" (*shufu*), unmodified, and thence to the "professional housewife" defined in part by well-honed skills in managing and fabricating the family wardrobe. These decades were likewise a time when the modern ideology of "good wife and wise mother," first articulated at the turn of the twentieth century, lived on in what Kathleen Uno has nicely characterized as "transmuted form."[76]

In terms of dress, this transwar era ended with the ascendance of ready-made women's wear. Articles began to appear in the press at the end of the 1960s lamenting that sewing machines "were sleeping in the house" as "it made more sense to buy than to make."[77] Time studies confirm that the practice of women sewing at home for the family had declined sharply by 1970.[78] And from an all-time peak in 1969, the domestic sale of sewing machines began a long, slow decline. In the realm of cooking, a similar shift took place toward reliance on prepared foods, although it took place later and more gradually. Most married women, including the increasing number who worked outside the home, retained the role of household manager, but the substance of this role changed. They spent less time fabricating or processing what the family consumed and more time working as purchasing agents who managed the acquisition of all manner of goods and services. In family and daily life, these trends together mark the end of the transwar era. More generally, this sense of an ending meshes with the interpretative stance of the sociologist Yoshimi Shunya, who argues that once the basics of the "bright new" consuming life were widely possessed in the 1970s, Japanese society entered a "post-postwar" era.[79]

Even so, in comparative perspective the professional housewife has had a remarkably long "post-postwar" afterlife. Japan throughout the second half of the twentieth century and into the early twenty-first ranked just after South Korea at the top of the industrial capitalist world in the depth of its "M-curve," the graphic representation of the proportion of women who leave the job market during years of childbirth and child-rearing.[80] In explaining this long endurance of the role of the full-time, or "majority-time," housewife, both as a social ideal and in lived experience, it is important to note negative incentives, such as an income tax system that penalized families with two wage earners unless the secondary earner worked part time. But to understand fully the dynamics of this endurance, one must recognize the importance of the positive social and cultural framing of the job of the skilled professional housewife, including her management of the family wardrobe. She was by no means created or validated only by the sewing machine and its associated skills. The *mishin* was one of several goods, and sewing was one of several realms of endeavor that brought a wider economy and culture intimately into the home. It provided a powerful connection to that wider world and affirmed her disciplined but optimistic role in managing those links.

Conclusion

A book focused on a single place inevitably invites the expectation of a singular story. What was "Japanese" about sewing machines in Japan? That question was raised in one way or another on almost every occasion when I described this project in the making. I conclude with two answers. Not as much as the question seems to hope or imply. And what is more interesting and important is the way in which so many of those elements that do seem particular to Japan were not so much made in Japan as forged by many hands in global interactions. The circulation of images, words, and the good itself across borders yielded practices of selling, buying, and using in which global and local were inseparable.

SEWING MACHINES AND A GLOBAL MODERN LIFE

Claims for Singer as "the world's first successful multinational company," and for sewing machines as "consumer durables which set the paradigm of mass production and consumption" are well founded.[1] These machines were indeed carriers and exemplars of a modernity and modernization that overspread the globe. Both Singer and its product played important roles for close to a century in propelling America's "irresistible" market empire to its dominant place not only in Europe but around the globe.[2] Our conclusions must therefore begin not with the phrase "in Japan in particular," but "in Japan as elsewhere."

In Japan, as elsewhere, the decades around the turn of the twentieth century witnessed the birth of the salesman as a practitioner of science and system, "the publicist for the happy sound of civilization" who convinced people to "replace carts with bicycles, pencils with typewriters, (oil) lamps with electric lights, and needles with sewing machines."[3] Not just any light bulb or sewing machine, but those mass produced with distinct brand names, whether GE or Matsushita, Singer or Janome. To be sure, these brands were cast in national roles, whether in "buy Japanese" campaigns of the late 1920s and 1930s or in the warlike rhetoric of trade competition, which persisted long after the war ended. But branding was ubiquitous, and the trade talk was itself virtually identical on both sides of the Pacific. And even as patriotic words poured forth, Janome's 1937 manual for sellers was titled "Our Salesmanship" in English on its cover.[4] Two decades later the company president called his salesmen "the most valuable asset in our company," and he clinched the argument by claiming that "99 percent of the presidents of the major companies in America started out in sales."[5]

With the birth of the salesman came the birth of the consumer, in Japan as around the world. This is not to be confused with the origins of consumption, whose worldwide history reaches back centuries, including its much-studied proliferation in early modern times in the consumption of all manner of goods by more people than ever.[6] What was new in the era of Singer as a global force was a consciousness of the "consumer" as a discrete figure not only in economic thinking but in society, politics, and culture. He or she—and her gender identity was problematized in Japan as around the globe—increasingly figured as a leading actor not only in getting and spending for herself or her family, but in making decisions that had to be wise, in the views of many, since they not only satisfied the wants of an individual but influenced the wealth of a nation and the fate of an empire. The sewing pedagogue Watanabe Shigeru, like the Singer salesman Endō Masajirō, lamented that too many women bought sewing machines merely to keep up with the neighbors. They wasted precious yen on a stupid purchase that went to little use. Truly "civilized" women would have invested in government bonds. That they bought *mishin* instead was a "major national problem."[7] In Endō's eyes, this only made the American company wealthy: as they sit unused, "sewing machines will do nothing for our country."[8]

But such talk did not mean that Japan offered a peculiarly hostile environment for the coming-of-age of the consumer. Even Watanabe admitted that, properly used, the sewing machine was a sensible purchase. The more practical Endō decided not to fight the spread of the *mishin* but to teach women to use them, and he founded what would became Japan's most successful sewing school. And when women wrote for each other about buying and using new goods, they were neither defensive or apologetic, nor worried about wasting the nation's wealth. In their telling, to buy a sewing machine was an eminently rational and modern act. Imura Nobuko urged in the magazine *Fujokai* in 1920 that women "increase the efficiency of the housewife or servant by boldly switching over to Western clothes for children."[9] She went on to urge her compatriots to buy *mishin* "with a small monthly installment payment, not that difficult for an ordinary family."

Indeed, credit in the form of the weekly or monthly installment loan was another innovation of the market empire pioneered worldwide by Singer and other firms from the late nineteenth into the early twentieth century, with some variation from place to place in discursive framing, though little difference in economic practice. It everywhere raised fears that weak-willed consumers would take on ruinous debt, fears that were countered as effectively in Japan as elsewhere, and possibly more so, by claims from the academic and business worlds that installment plans would induce consumers to "lead a disciplined life, planning a monthly budget of expenses."[10]

In addition, the high value that discourse surrounding the sewing machine placed on its potential to help women achieve a measure of independence or self-sufficiency, "just in case" they had to support themselves, was not a particular Japanese adaptation of a global female modernity. It was a value put forth elsewhere as well, which has not been sufficiently recognized as part of the thinking of and about women in Japanese modern times.

In the cultural life of the consuming subject or citizen, one sees in Japan as elsewhere a two-sided modernity projected at and anchored in the imagination of a new middle class. In prewar years this was a minority group whose quality of life the majority could only dream of obtaining, while those who possessed it were always anxious about losing it. An ad such as Janome's of 1936 (figure 20) appealed to these dreams by invoking Americans as the model for women in Japan, who were told to buy a scientific product—the sewing machine—so they could economize on time and expenses *and* use their profits to indulge with abandon in the joy of shopping.[11] Similarly, if a bit less flamboyantly in France, in Judith Coffin's thoughtful reading of print advertising, "The sewing machine had a dual meaning; it was a mark of a modern household and of access to 'the world of goods,' but it also represented hard-nosed calculations about family incomes without which such access would be impossible."[12] It is tempting to see the Japanese culture of modernity, with its two quite different words to express the concept of "modern," as particularly explicit in simultaneously bringing to the foreground both the soberly future-oriented modern (*kindai*) life of rationality, self-discipline, and earning or saving for investment, and the stylishly present-focused modern (*modan*) life of play and pleasure. Perhaps this linguistic splitting of the English word *modern* is unusual; but if so, the Japanese language helps us to understand not something unique to that place, but that the coexistence of these dual aspects is the essence of the modern condition everyplace.

In its social life, this good was also Janus-faced. It could be a tool of oppression for exploited homeworkers, but it validated the pursuit of self-sufficiency by women in Japan, and it offered them some limited, but not unattainable, opportunities. This was certainly the case around the world as well. One is hard-pressed to identify places in the era of machine sewing where women in significant numbers did not undertake both family sewing and home-based commercial sewing. Barbara Burman's claim for North America and Britain from the mid-nineteenth through the mid-twentieth century certainly applied to Japan from early in the twentieth century through the 1960s, arguably to a greater extent: women commonly took on the "unwaged making of clothing in the home," and "many of the same women have reapplied the same variety of skills to sew for a network of neighbours or workmates for money or taken in sewing as waged piece-work on a regular basis."[13] Only as masses of consumers came to prefer industrially produced, ready-to-wear garments (for whatever mix of motives and trade-offs involving cost and convenience, style and quality), did home-based commercial dressmaking recede as a woman's ideal or practice.

In a world where the boundary between sewing for wages and sewing for self and family was permeable and often traversed, fabricating and managing the family wardrobe was one defining responsibility of the full-time, or what I have called for Japan, the professional housewife. Here—as suggested in chapter 7—I do see local variation on this global theme that is worth noting. But it is important to note that even as she was portrayed with particular national or colonial characteristics, the image of the modern woman as a wife and mother whose primary vocation was to manage the home was powerfully reinforced, and in some measure invented, through the global marketing of the tools of this woman's trade, the sewing machine prominent among them. Strong visual evidence of this process of marketing the modern family is found in Singer's "trans-

lations" of an early twentieth-century American, or Euro-American, bourgeois family ideal into a variety of non-Western cultural settings. In postcards that circulated in the United States, such family tableaux demonstrated the worldwide appeal of Singer's machine, which one assumes
boosted the appeal of the product "at home."[14] In addition, and of more
importance to us, as this translated family scene circulated in Japan, it
gave legitimacy to the new and modern ideal of the good wife and wise
mother, even as it partially disguised her Western origins.

The family in Singer's advertising, both in its original American depiction and in translation in Japan, or Java, or Burma, stood clearly at
the comfortable upper end of the modern middle class. Imura Nobuko,
as she endorsed Western machine sewing for its efficiency, expected some
readers to be sufficiently comfortable to have domestic servants. But with
the phrase "increase the efficiency of the housewife or servant" she was
aiming as well at a broad pool of readers.[15] One argument of this study
has been that the sewing machine held a broad appeal across social classes
in Japan. Until the 1950s, only a minority could afford to buy this good.
One can imagine that in previous decades a sense of exclusion from the
world of the middle class, with its exciting new possessions, would have
only heightened the resentment of working or lower-class women and
families. But in Japan it seems that the widespread desire for this good,
even if it outran the possibility of buying it, served more to bridge than
to deepen a divide of social class. Matsunami Tadayuki claimed on behalf of consumer credit circa 1930 that it would "democratize mass access to commodities, spread human happiness more equally, and relieve
troubling class struggles.[16] His rhetoric was grandiose, but it captured a
kernel of truth, echoed as well in the way women who used their sewing
machines for economic survival nonetheless found some solace or attraction in their vicarious participation in broader cultural and fashion trends.

Historians have argued that in Europe and America the sewing machine turned working-class households "into miniature sweat shops"
while it "had a different effect" on middle-class women who sewed for
their families.[17] For the Victorian-era middle class, that is, the sewing
machine "symbolized leisure and conspicuous consumption," while for
women of the working class it "signified the exploitative conditions of
the factory and homework."[18] If such a view is upheld, the integrative impact of the sewing machine appears particular to Japan or at least divergent from Western experience. But more recent studies have put forth a
different understanding of the American and British cases. Fiona Hackney persuasively concludes that dressmaking as promoted in magazines

for working-class women in interwar Britain "signified a sense of belonging, and of participating in the new commercial cultures of fashion and modern femininity."[19] Nancy Page Ferguson echoes this claim for the United States of the late nineteenth century, where sewing machine promoters "addressed anxieties about the social impact of industrial change." She concludes that while sewing machine advertising presented "multiple narratives" that indeed acknowledged divergent gender and class identities, these marketing efforts in sum "disassociated consumption from economic change and relieved anxieties over the impact of industrialization on American life."[20] In Japan as well, I believe the sewing machine, among other goods, affirmed social order and bridged class differences more than it provoked conflict or disorder; it validated a significant range of ideals and roles for women in the modern world, from the seeker of "cultivation" and the happy participant in the global swirl of styles and fashion to the disciplined and skilled consumer and manager of the home, and the homeworker proud to be self-reliant.

RESISTING AND ADAPTING THE MODERN

To emphasize the manifold ways in which consumers in Japan spoke a language and followed a path of a modern life that circled the globe from the mid-nineteenth century is important. But this is neither to claim that these messages and meanings emanated continuously from one source moving in one direction, nor to assert they were absorbed without challenge or change. Globalizing is indeed a localizing process, marked by variations from place to place on the large themes outlined above. Aspects of modern culture and economic life had some particular points of origin, but their spread was multidirectional. They touched down here and there, circulated in many directions, and even looped back to where they began. The scientific management of industrial production and its later incarnation in what Americans called statistical quality control were significantly adapted in Japan of the 1950s and 60s and then cycled back to the United States and around the world in the new form of a movement for quality-control circles.[21] The modern girl of the interwar era, as brought to life in a brilliant recent study, likewise was the product of what the authors persuasively describe as an asymmetrical process of "multi-directional citation."[22] She appeared Caucasian in cosmetic ads in Japan, yet distinctly Asian in, of all places, Germany in the very years that the Nazi party rose to power. In the postwar world of fashion—building, it seems certain, on the foundation provided by a nation of savvy dress-makers and schools of clothing design—Japan emerged as not only re-cipient but producer of globally desired fashions by figures such as Mori Hanae, Issei Miyake, or Kawakubo Rei. Even while they simultaneously tweaked the Singer system and carried it forward within Japan, in the production and sale of sewing machines for export to the United States in the 1950s and 1960s, Japanese companies and their American partners improvised strategies of mass retailing that "mighty Singer" was itself unwilling to adopt.

As the practices of the market empire crossed the Pacific beginning around the turn of the twentieth century, they were met with a charged rhetoric of resistance. Episodes such as the resistance by the salesmen (and then by new competitors) to Singer in the 1930s, the efforts to reform the nation's dress during wartime, or the struggle to prevent Singer's post-war return generated a self-understanding of difference among nations or cultures, or of clashing empires, that outran the actual differences in how goods were sold or in how people dressed or enacted their lives. In these episodes, the endeavor to articulate a supposedly Japanese variant of modern practices was more important and consequential than any out-right denial of or resistance to a global modernity.

One sees a Japanese-style modernity invoked in the debates over the double, or "two-layered," life of the 1920s, in which participants strove to identify and promote a life that was in some sense "Japanese" even as it accepted or embraced modern values of speed and efficiency, freedom of movement, and sometimes even freedom of choice. One sees it artic-ulated as well in feverish attempts in the era of wartime modernity to de-vise "national" and "standard" dress that was both rational in its home-front suitability for the war effort and a uniquely "Japanese outfit that can play a leading role in world clothing culture."[23] At a general level these endeavors were hardly particular. People in both Japan and Amer-ica "shaped cultural constructions of the family sewing machine" and new habits of dress in an effort to cope with "anxieties about the impact of industrialization" and to make sense of the disorienting changes of modern life.[24] But there is an extra or doubled element in the Japanese case, as elsewhere in the non-West: those who endeavored to cope with change understood it not so much in temporal terms, for instance, as pre-serving or reshaping older traditions in a newer age, but in spatial terms, as the effort to cope with a culturally foreign intrusion that brought both possibility and danger.[25]

If in some cases these efforts took the form of struggle and resistance, the result was more adaptation or inflection than rejection. Salesmen and store managers struggled with Singer's expatriate executives over the terms of their employment. Similar demands were raised in disputes of the same era by workers in companies wholly owned and managed by Japanese capitalists, but the Singer employees claimed to be resisting "Yankee capitalism," and their battles took on particular intensity because of that framing. The fight with Singer deepened the understanding of systems of retirement and severance pay as an "appropriately Japanese" variant of capitalism. These variations fed into the institutional complex that came to be understood as the "Japanese employment system," and by the postwar era, in both practice and understanding, this was indeed a significant variation on the theme of employment in a capitalist system.

The Japanese competitors who began to erode Singer's dominant position in the mid-1930s described their credit system as derived from indigenous precedents, likewise more appropriate to Japan's economic situation, and their innovations in selling and in compensating canvassers with base pay plus commission were significant. But in the end these were not so much frontal assaults on Singer practices as modest adaptations, for which precedents existed in the United States as well. Into the early postwar decades, especially in the 1950s, Janome and other competitors spoke of resisting an invasion by American industry and defending ethnic industry. Their rhetoric of warring nations evoked the recent fury of the war of empires. Janome also gave its selling practices an indigenous gloss with the idiom *deep cultivation*. The intensity of this trade-war talk is noteworthy; it complicates a narrative of the sudden evaporation of wartime antagonism in the amity of an era described by many, and not only those on the left, as one of "peace and culture." But the idea of "deep cultivation" notwithstanding, it would be a mistake to take this rhetoric as a sign of a fundamentally divergent business paradigm. Janome was engaged in business competition among various modern corporations, some domestic, some foreign, who made the same products and sold them with a modified version of the Singer selling system.

Turning to the use of these machines, some important adaptations do come into relief as the nation recovered and prospered, and as the middle-class family and the professional housewife rose to cultural and social prominence. The fact that she so often enrolled in commercial sewing classes distinguishes the Japanese professional-housewife-in-training from her counterparts. Some learned needle skills at home from a mother, or siblings, or in the standard home economics curriculum of the public schools, without post-graduate schooling, but informal training was much less prevalent than in Britain or the United States, and the very common British practice of apprenticeship to a skilled dressmaker was also rare.[26] The Japanese pattern of formal sewing education in commercial academies reflected and furthered the belief that women home managers carried a specialized set of professional skills.

Related to this well-honed sense that the housewife's job was a specialized and credentialed vocation—and one element that appears to stand as a distinctive cultural construction—is the widespread view that women, more than men, were responsible borrowers and shoppers. It may be that the image of the female consumer as rational is not so much unique to Japan as uniquely emphasized, both by contemporaries and later by scholars, and under-recognized in other places. This is a topic worthy of further study. But I find the figure of the woman managing home finance with professional attention, struggling to rein in her spendthrift husband's binging on credit tickets, to be a singular one, linked to the singularly enduring influence in high-growth, postwar Japan of the ideal of the professional housewife.

Finally, although women virtually everywhere have sewn at home for their families and for wages, home-based sewing of all sorts appears to have been especially preponderant in Japan. This was already the case well before the advent of the sewing machine, at least in comparison to European or American societies, where professional dressmaking earlier and more extensively moved outside the home and toward the industrial production of garments. At the start of the twentieth century, the sewing machine entered Japan's world of home-based sewing and dressmaking, for the first few decades in modest numbers in global comparison, but always with a strong emphasis on family machines and home users, and with a broader cultural impact as the desire grew to own this symbol of modernity. By the interwar years, commercial schools, print media, and both civilian and state reformers had begun to piece together the complex of practices that made the middle- or working-class mother or daughter a remarkably frequent needleworker even in her "free" time. Wartime policies built upon and reinforced the ubiquity of home sewing. The state and allies such as Narita Jun mobilized the needleworking housewife as a strategic resource and promoted "national" and "standard" dress that was amenable to home sewing, whether by hand or by machine. In contrast to other combatant nations, both scarcity and the relatively slow mobilization of young women into factories meant that no significant countervailing trend emerged of wage-earning women turning to ready-made garb. Accustomed to remarkably long hours of family sewing, women of all classes and regions stepped out of the bleakness of a nonetheless modern wartime landscape into the bright new life of postwar times, and the interwar complex of sewing promotion flourished as never before. Women trained in burgeoning commercial schools. They practiced their skills for profit or for family, for survival or for leisure. Sewing at home for as many as two or three hours a day, every day, these women helped make Japan for more than two decades, and, with a long afterlife, a nation of both professional housewives and fabricating consumers.

Some Notes on Time-Use Studies

Social surveys that documented the long hours of sewing among Japanese women in the 1950s sparked my initial interest in the questions that led me to write this book. It thus seems appropriate to offer some additional information and analysis that does not fit neatly into the exposition of the book itself, but is nonetheless of some relevance to this project.

The social scientific study of time use in daily life began in the years after World War I. State agencies and some private organizations in the United States and Great Britain, as well as the state in the Soviet Union, were pioneers in this field, but Japanese officials were also among the earliest to undertake survey research of time use, with the Osaka city government's study of "free time use" noted in chapter 3 as the local pioneer.[1] The introductory chapter of *Time Use Research in the Social Sciences* offers a useful survey of the historical development of time studies.[2] These early efforts were undertaken to provide data that might help worried state officials and others address the "social questions" of the day, such as the habits of those in the working class who were feared to lack the discipline or work ethic needed to insure social order. Such concerns clearly motivated the first study in Osaka as well as the studies by various agencies in Japan in the 1950s.

Additional factors motivating the early endeavors in this field included— during World War II—the concern to mobilize entire populations as efficiently as possible in the cause of total war and, beginning with the age of radio, the desire of broadcasters to target the right audiences at the right times by pinning down the daily routines of listeners (and later, television viewers). The BBC from the late 1930s was the pioneer in Britain and globally in this effort to understand the daily flow of household time use, and NHK in Japan undertook its own ambitious surveys just a few years later. After World War II, the most important cross-national collaborative and comparative studies grew out of a conference held at Yale University in the early 1960s, which resulted in the publication of a major volume a decade later.[3]

In this book, I have used the time studies for two purposes: to get a sense of the hours that women in Japan spent at home-based sewing for themselves and their families, and to estimate how many women undertook homework (*naishoku*). In both cases, one question worth discussion is that of the meaning of an average number of minutes spent on either family sewing or commercial, home-based sewing. That is, if women averaged a certain number of minutes per day at their sewing machines, might that total not mask a division into groups of non-sewers and frequent sewers? What is the distribution around the mean?

Looking first at noncommercial family sewing, in the Japanese case, unfortunately, only the 1970 iteration of NHK's time-use surveys provides data that address these questions. In a perspective unfortunately not presented in NHK's earlier studies or those of other agencies, this 1970 survey reported that, on average, among all women, sewing occupied thirty-three minutes each day, while full-time household managers spent about an hour. Concerned to get a sense of the distribution around the mean, however, the survey then set aside all those who reported they sewed not one stitch. It turned out that those who did any regular sewing at all devoted a substantial amount of time to it: an average of two hours and sixteen minutes. The averages—which had declined sharply from the 1950s—thus masked a division among women into camps of frequent sewers and non-sewers.[4]

The same bifurcation may have existed earlier without having been calculated or reported for posterity, and of course it is always true to some extent that distribution exists around any mean accounting of social behavior. Still, I think it is fair to conclude that the higher the average daily numbers of sewing hours and minutes, the less wide was the spread around the mean. That is, when full-time housewives in Japanese homes were found to be averaging two or even three hours of *noncommercial* needlework at home in the 1940s and 1950s, it is unlikely, given that they were also cooking, shopping, cleaning, and usually doing some childcare, that many of them were doing family sewing for, say, five or six hours each day; conversely, a daily average of two to three hours of sewing probably did not include a large proportion who did no sewing at all.

The matter is different when we turn to commercial homework. Here it is certain that any average report of homeworker minutes derived from a survey that included those who did homework and those who did not is of limited value without some reworking of the aggregated data. In chapter 5 I proceed as follows to derive what I believe is a conservative estimate that wartime homework (at the time of the 1942 survey) was undertaken by women in 3 percent of households headed by a middle-class office worker (salaryman) and in 13 percent of households headed by a male factory worker. The NHK surveyors had produced their average of time spent at such homework by simply adding up the total time devoted to homework and dividing it by the total number of people surveyed. That produced averages of about ten minutes per day devoted to homework by the 2,040 women in salaryman families, and forty-five minutes per day for the 1,776 women in families of male factory laborers.[5] Since commercial homework, unlike cooking, family sewing, or child-care, was only undertaken by a minority of households, these averages are not very meaningful except to suggest that the relative frequency of homework in working-class families was a good bit more than in the middle class. Early postwar surveys that looked only at those who *did* engage in homework can be used to adjust the data. These found that homeworkers spent an average of 4–6 hours per day at this task. If we assume a similar average obtained a decade earlier (and take six hours as the average to yield a conservative estimate), we can rework the data as follows: the surveyed women in middle-class (salaryman) families devoted a total of 20,400 daily minutes to homework. Dividing by an estimated 360 minutes spent by each homeworker yields fifty-seven individual homeworkers, 3 percent of the total number of women surveyed in this category. Applying this procedure to the factory workers yields an estimate that 13 percent of their households had a wife or daughter engaged in homework (1,776 women × 45 minutes ÷ 360 = 222 homeworkers, and 222 ÷ 1,776 = 12.5 percent).

Notes

PREFACE

1. On the "special resonance and potency" of home sewing in Britain and North America, see Barbara Burman, "Introduction," in *The Culture of Sewing: Gender, Consumption and Home Dressmaking,* ed. Barbara Burman (Oxford: Berg, 1999), p. 5.

2. Letter to author, from Professor Kinoshita Jun, April 23, 2003.

3. Here and throughout the book, Japanese names are written as in Japan: the family name precedes the given name. Names of Japanese American scholars are written with given name first.

INTRODUCTION

1. Kawasaki rōdō shi hensan iinkai, ed., *Kawasaki rōdō shi,* p. 200, citing a March 1950 survey prepared by Rōdō igaku shinrigaku kenkyūjo.

2. I am much indebted to Louise Young for suggesting this.

3. The survey was conducted by the Ministry of Labor, Women and Minors Section. Rōdō shō, fujin shōnen kyoku, *Kōjō rōdōsha seikatsu no chōsa* (Tokyo: Rōdō shō, fujin shōnen kyoku, 1952), p. 85 for the time-use data; p. 36 for machine ownership rates.

4. Karl Marx, *Capital* (New York: Charles H. Kerr, 1921), vol. 1, pp. 516–18.

5. Mohandas Gandhi, *All Men Are Brothers: Life and Thoughts of Mahatma Gandhi as Told in His Own Words,* ch. 7, "Man and Machine," www.mkgandhi .org/amabrothers/chap07.htm (accessed 3/14/2011).

6. Marx, *Communist Manifesto* (Chicago: Charles H. Kerr, 1906), p. 17.

7. Gandhi, *All Men Are Brothers,* ch. 7.

8. Victoria de Grazia, *Irresistible Empire: America's Advance through Twen-*

tieth Century Europe (Cambridge, MA: Harvard University Press, 2005), pp. 7–8, 205–9.

9. Arjun Appadurai, *Modernity at Large: Cultural Dimensions of Globalization* (Minneapolis: University of Minnesota Press, 1996), pp. 17, 32. See also C. A. Bayly, *The Birth of the Modern World, 1780–1914* (Oxford: Blackwell, 2004), pp. 1–2.

10. Film version of *Madame Butterfly*, directed by Frédéric Mitterrand (Columbia Tristar Home Video, 1997).

11. Tim Putnam, "The Sewing Machine Comes Home," pp. 269–70; see also Nicholas Oddy, "Beautiful Ornament in the Parlour or Boudoir: The Domestication of the Sewing Machine," p. 295, both in Burman, *Culture of Sewing*, on its impact on the design of numerous other appliances.

12. Gaimushō kiroku, "Honpō ni okeru rōdō sōgi kankei zakken: 'Shingā mishin' kaisha kankei," p. 255 ("Statement" submitted to T. Taketomi, in charge of Foreign Trade, Foreign Office, Kasumigaseki, Tokyo, 11/12/1932). Held in National Archives of Japan, Japan Center for Asian Historical Records, Tokyo.

13. Penelope Francks, "Inconspicuous Consumption: Sake, Beer, and the Birth of the Consumer in Japan," *Journal of Asian Studies* 68, no. 1 (February 2009): 135–60.

1. MEIJI MACHINES

1. Donald R. Bernard, *The Life and Times of John Manjiro* (New York: McGraw Hill, 1992), and Christopher Benfey, *The Great Wave: Gilded Age Misfits, Japanese Eccentrics, and the Opening of Old Japan* (New York: Random House, 2003); ch. 1 narrates Manjiro's tale.

2. "They Inspect Wheeler and Wilson's Sewing Machine," *Frank Leslie's Illustrated Newspaper*, June 9, 1860, p. 27.

3. Bernard, *John Manjiro*, p. 199.

4. Yoshida Gen, "Nihon saihō mishin shi zakkō," *Mishin sangyō*, 100 (1967): pp. 1–10.

5. Bernard, *John Manjiro*, p. 128.

6. Both ads are held in the Smithsonian Archives Center, Warshaw Collection, "Sewing Machines," box 5, folder 2.

7. Benfey, *Great Wave*, pp. 39–40, cites the diary of Francis Hall, who records a conversation with Manjiro about his mother's reaction.

8. Liza Crihfield Dalby, *Kimono: Fashioning Culture* (New Haven, CT: Yale University Press, 1993), p. 10.

9. Nakayama Chiyo, *Nihon fujin yōsō shi* (Tokyo: Yoshikawa kōbunkan, 1897), pp. 298–323.

10. C. A. Bayly, *The Birth of the Modern World, 1780–1914* (Oxford: Blackwell, 2004), pp. 12–13.

11. Basil Hall Chamberlain, *Things Japanese* (London: K. Paul, Trench, Trubner, 1891), p. 1.

12. Robert Bruce Davies, *Peacefully Working to Conquer the World: Singer Sewing Machines in Foreign Markets, 1854–1920* (New York: Arno Press, 1976), ch. 4.

13. Nira Wickramasinghe, "The Reception of the Singer Sewing Machine in Colonial Ceylon/Sri Lanka," unpublished paper presented at Princeton University, Davis Center, March 27, 2009, p. 20.

14. For sales data, see "Notes on China and Japan," June 1, 1884, and undated "Summary of Business, China and Japan 1882–1886." On Sang's alleged theft, see "Letter from Mitchell to London Office," June 1, 1888. All three items held in Wisconsin State Historical Society, Madison, WI, Singer Sewing Machine Company Collection, call no. US Mss A1.

15. Nihon mishin kyōkai, ed., *Nihon mishin sangyō shi* (Tokyo: Nihon mishin kyōkai, 1961), p. 24.

16. The characters are *sai-dōgu*. See Janome mishin shashi hensan iinkai, ed., *Janome mishin sōgyō 50 nen shi* (Tokyo: Tokiwa shoin, 1971), p. 92.

17. An ad using the word *mishin* first appears in the *Kyoto shinbun* in November 1872. See Meiji hennenshi hensankai, ed., *Shinbun shūsei Meiji hennenshi*, vol. 1 (Tokyo: Zaisei keizai gakkai, 1934), p. 526. The first ad in the *Yomiuri shinbun* appeared on August 23, 1875, p. 2. On Asakusa, see Shimokawa Kōshi, ed., *Meiji Taishō katei shi nenpyō* (Tokyo: Kawade shobō shinsha, 2000), p. 44. In the early days of the sewing machine, demonstrations to paying customers were common around the world.

18. "Present from Mr. Roosevelt to the Empress of Japan," *Times* (London), September 25, 1905, p. 4.

19. Sally A. Hastings, "The Empress's New Clothes and Japanese Women, 1868–1912," *Historian* 55, no. 4 (Summer 1993): pp. 681–82.

20. The empress's circular was published in the first issue of *Nihon no jogaku,* January 1887, and according to Hastings it also appears in *Chōya shinbun,* January 17, 1887.

21. Hastings, "Empress's New Clothes," p. 682.

22. Ibid.

23. The translation of the circular is my slightly modified version of a portion from Hastings, "Empress's New Clothes," p. 683.

24. Sharon H. Nolte and Sally Ann Hastings, "The Meiji State's Policy toward Women," in *Recreating Japanese Women: 1600–1945,* ed. Gail Lee Bernstein (Berkeley and Los Angeles: University of California Press, 1991), p. 152.

25. Ibid., p. 168.

26. Ibid., p. 171.

27. "Shingā saihō jogakkō, shiritsu gakkō setsuritsu onegai" (9/16/1906). Gakuji kankei kenmei mokuroku (Tokyo fu, Tokyo shi). Held at Tokyo Metropolitan Public Records Office.

28. Hastings, "Empress's New Clothes" pp. 689–92.

29. Carol Gluck, *Japan's Modern Myths: Ideology in the Late Meiji Period* (Princeton, NJ: Princeton University Press, 1985), p. 137.

30. Abe Tsugio, ed., *Gyōkai konjaku monogatari* (Tokyo: Nihon mishin taimusu sha, 1960), p. 25 (transcript of a discussion by veterans of the earliest days of the industry in Japan convened in November 1947).

31. The flyer is held at the Nature and Science Museum of the Tokyo University of Agriculture and Technology. The manual, "Shingā saihō kikai shiyō hō, dai 15, 27, 28 shū," is available in the National Diet Library.

32. This is the speculation of Kobayashi Norio, president of the Japan Antique Sewing Machine Society and volunteer curator of the sewing machine collection at the Nature and Science Museum of the Tokyo University of Agriculture and Technology (interview on March 11, 2003).

33. Hata Toshiyuki, "Shingā seizō gaisha ni kan suru hōkoku" (1903), pp. 3–4; reprinted in Matsumura Satoshi, *Kaigai jitsugyō renshū sei hōkoku, Nōshōmushō shōkōkyoku rinji hōkoku,* vol. 10 (Tokyo: Yumani shobō, 2002), pp. 325–28.

34. On Singer's claim that its machine "elevates human nature" worldwide, see Davies, *Peacefully Working to Conquer the World,* pp. 97–98. Mona Domosh further describes this globally broadcast message of a civilizing mission in *American Commodities in an Age of Empire* (New York: Routledge, 2006), ch. 1.

35. Anne Hollander, *Sex and Suits* (New York: Albert A. Knopf, 1994), p. 117.

36. Ibid.

37. Andrew Godley, "Homework and Sewing Machine in the British Clothing Industry, 1850–1905," in Burman, *Culture of Sewing,* p. 260.

38. Wendy Gamber, *The Female Economy: The Millinery and Dressmaking Trades, 1860–1930* (Urbana: University of Illinois Press, 1997), pp. 8, 5 respectively for the two quotations.

39. Ibid., p. 216.

40. Antonia Finnane, in *Changing Clothes in China: Fashion, History, Nation* (New York: Columbia University Press, 2008), writes similarly of twentieth-century China: "Women as professional tailors or dressmakers were latecomers; . . . the millions of women at home made the greater bulk of clothing" (p. 114).

41. Dalby, *Kimono,* p. 70.

42. Kären Wigen, *A Malleable Map: Geographies of Restoration in Central Japan, 1600–1912* (Berkeley and Los Angeles: University of California Press, 2010), p. 162.

43. Dalby, *Kimono,* p. 70. The phrase in brackets is my insertion.

44. Ibid., p. 83.

45. Nakayama, *Nihon fujin yōsō shi,* pp. 306–22. See also Yōfuku kisha kurabu, ed., *Nihon no yōfuku shi* (Tokyo: Yōfuku kisha kurabu, 1976), pp. 58–111.

46. Nakayama, *Nihon fujin yōsō shi,* pp. 322–28.

2. THE AMERICAN WAY OF SELLING

1. Geoffrey G. Jones and David Kiron, "Singer Sewing Machine Company: 1851–1914," Harvard Business School Case N9–804–001 (Boston: Harvard Business School Publishing, 2003), p. 1.

2. On Singer's early history, see Fred V. Carstensen, *American Enterprise in Foreign Markets: Singer and International Harvester in Imperial Russia* (Chapel Hill: University of North Carolina Press, 1984), pp. 17–19, 24–25. See also Andrew Godley, "Selling the Sewing Machine Around the World: Singer's International Marketing Strategies, 1850–1920," *Enterprise and Society* 7, no. 2 (June

2006): pp. 270–76, 302; Robert Bruce Davies *Peacefully Working to Conquer the World: Singer Sewing Machines in Foreign Markets, 1854–1920* (New York: Arno Press, 1976), p. 161.

3. Godley, "Selling the Sewing Machine," pp. 268–69.

4. It only held that title for one year, surpassed by the Metropolitan Life building. But in 1967 it became the tallest skyscraper ever to have been torn down. One Liberty Plaza sits on the site today.

5. These phrases appear at the head of a Singer Japanese-language sales pamphlet, "Katei oyobi shokugyō yō shingā mishin mokuroku," in the collection of the Edo-Tokyo Museum, item no. 93200196, undated.

6. Nihon mishin kyōkai, ed., *Nihon mishin sangyō shi* (Tokyo: Nihon mishin kyōkai, 1961), p. 24.

7. Kuwahara Tetsuya, "Shoki takokuseki kigyō no tainichi tōshi to minzoku kigyō," *Kokumin keizai zasshi* 185, no. 5 (May 2002): p. 48.

8. Hata Toshiyuki, "Shingā seizō kaisha ni kan suru hōkoku," p. 2; reprinted in Matsumura Satoshi, *Kaigai jitsugyō kenshūsei hōkoku, Nōshōmushō shōkō-kyoku rinji hōkoku,*" vol. 10 (Tokyo: Yumani shobō, 2002), p. 326.

9. Godley, "Selling the Sewing Machine," p. 282, draws on Walter Friedman, *The Birth of the Salesman: The Transformation of Selling in America* (Cambridge, MA: Harvard University Press, 2004).

10. Godley, "Selling the Sewing Machine," p. 291.

11. Ibid., p. 296.

12. Hata, "Shingā seizō kaisha," p. 3.

13. "Directory of Shops under Controlling Agency [1904?] and "Directory of Shops for the Sale of Singer Sewing Machines Throughout the World, Revised, January 1906," both in the Singer Sewing Machine Collection in the Wisconsin State Historical Society.

14. Osaka Trade and Industry Bureau, ed., *Mishin kōgyō* (Osaka: Mishin geppō sha, 1951), p. 315.

15. Abe Tsugio, ed., *Gyōkai konjaku monogatari* (Tokyo: Nihon mishin taimusu sha, 1960), p. 28 reprints a 1948 roundtable discussion.

16. Ibid., p. 29.

17. Ad headlined "Singer's Sewing Machines," undated, Smithsonian Institute Archives, Warshaw Collection, Sewing Machines, box 3, folder 1. The text notes that Singer quality "has just been confirmed" by a Gold Medal at the Great Exposition at the Palace of Industry, in France, placing it either shortly after the 1878 or the 1889 exposition.

18. Abe, *Gyōkai konjaku monogatari,* p. 28.

19. "Mishin gyōkai to seerusuman no kimyō na kankei," *Nihon shūhō,* April 15, 1964, p. 38.

20. Friedman, *Birth of a Salesman.*

21. Ibid., p. 88.

22. Ibid., p. 91.

23. Ibid., p. 97.

24. Gaimushō kiroku, "Honpō ni okeru rōdō sōgi kankei zakken: 'Shingā mishin' kaisha kankei," pp. 449–56 (Koyama Okazaemon, "Watakushi no itsu-warazaru kokuhaku"). Held in the National Archives of Japan, Japan Center for

234 I Notes to Chapter Two

Asian Historical Records. Hereafter this main source is cited simply as Gaimushō kiroku.

25. Shimozono Satoshi, *Dotō o koete: Yamamoto Tōsaku denki* (Tokyo: privately published, 1960), pp. 52–53. This hagiographic biography may exaggerate his ranking, but Yamamoto was without doubt a successful salesman.

26. Gaimushō kiroku, p. 590 (Matsuoka Komakichi, "Shingā kaisha sōgi no keika to sono shinsō," December 15, 1932).

27. Gaimushō kiroku, pp. 430–37 ("Yōkyusho").

28. For documentation of both these disputes, see Kyū kyōchōkai shiryō, *Rōdō sōgi,* vol. 9 (1925), held at the Ohara Institute for Social Research.

29. Gaimushō kiroku, pp. 322–23 (Gaihi dai 2349 gō, 11/18/1932).

30. Gaimushō kiroku, p. 249 ("Statement" to "Mr. T. Taketomi, in charge of Foreign Trade, Foreign Office"; original in English).

31. Gaimushō kiroku, pp. 875–76 ("Vest yori no kanyūjō," 1/11/1933).

32. Yoda Shintarō, *Gakuriteki shōryaku hō: Hanbaiin to hanbaijustu* (Tokyo: Hakubunkan, 1916). Taylor's *Principles* were introduced in the popular press in Japan almost immediately upon publication in 1911, and the book was published in a full translation in 1913. See William Tsutsui, *Manufacturing Ideology: Scientific Management in Twentieth-Century Japan* (Princeton, NJ: Princeton University Press, 1998), pp. 18–19.

33. Yoda, *Gakuriteki shōryaku hō,* p. 5.

34. Kuramoto Chōji, *Atarashii gaikōjutsu* (Tokyo: Seibundō shinkō sha, 1936), pp. 14–15, 17.

35. Ishikawa Rokuro, *Shusse gaikōjutsu* (Tokyo: Jitsugyō no nihonsha, 1925), pp. 6–7.

36. Shimizu Masami, *Hōmon hanbai chūmon o toru hiketsu* (Tokyo: Clark sōsho kankōkai, 1924), pp. 1–6; and *Shin gaikō hanbai jutsu* (Tokyo: Seibundō shinkō sha, 1937).

37. Gaimushō kiroku, p. 457 (Tokyo onna kyōshi bu, "Seigansho").

38. Hata Rimuko, *Mishin saihō hitori manabi,* 3rd ed. (Tokyo: Hata shoten shuppan bu, 1933), preface, p 1.

39. Gakuji kankei kenmei mokuroku (Tokyo fu, Tokyo shi), "Shingā mishin saihō jogakkō, shiritsu gakkō inchō shūshoku ninka," 10/16/1906, held in the Tokyo Metropolitan Public Records Office.

40. Gakuji kankei kenmei mokuroku, "Shingā mishin saihō jogakkō, shiritsu gakkō setsuritsu onegai," 9/16/1906, in the Tokyo Metropolitan Public Records Office.

41. Gakuji kankei kenmei mokuroku, "Shingā saihō in, gakkō setsuritsu no ken," 9/23/1919, in the Tokyo Metropolitan Public Records Office.

42. *Fujo shinbun,* November 3, 1906.

43. Ōue Shirō, ed., *Meiji kakō chō: Bukko jinmei jiten* (Tokyo: Tokyo bijutsu, 1971). Riu's life was celebrated in the 1936 Nikkatsu film, *Bokoku no haha.*

44. Jordan Sand, *House and Home in Modern Japan: Architecture, Domestic Space, and Bourgeois Culture, 1880–1930* (Cambridge, MA: Harvard Asia Center Monographs, 2003), p. 346.

45. Miyoshi Hisae, "Kodomo fuku no saihō mise wo hajimete," *Fujokai* 45, no. 2 (February 1932): pp. 282–84. This was one of three popular magazines of

the time whose titles translate as *Women's World*. The others were *Fujin sekai* and *Fujinkai*.

46. "Mishin no hibiki," *Tokyo Asahi shinbun,* June 22, 1932, p. 7.

47. Furuga Zansei, "Onna kyōshi no mondai (1–3)," *Tokyo Asahi shinbun,* May 20–22, 1935 (p. 9 of each issue).

48. Hata, "Shingā seizō kaisha," p. 5.

49. On "*kō,*" see Tetsuo Najita, *Ordinary Economies in Japan: A Historical Perspective, 1750–1950* (Berkeley and Los Angeles: University of California Press, 2009).

50. See *Geppu kenkyū* 1, no. 1 (April 15, 1957): p. 3; 1, no. 3 (June 15, 1957) p. 3; and Tokyo shōkō kaigisho, ed., *Geppu hanbai seido* (Tokyo: Tokyo shōkō kaigisho, 1929), pp. 212–13.

51. Lendol Calder, *Financing the American Dream* (Princeton, NJ: Princeton University Press: 1999), pp. 56–57.

52. On installment department store rates, see Kuribayashi Sho, "Geppu no hanashi," *Sarariiman* 2, no. 3 (March 1929): pp. 65–66.

53. "Annual World Reports," held at Singer Sewing Machine Collection, Wisconsin State Historical Society, Madison, WI.

54. "Katei oyobi shokugyō yō shingā mishin mokuroku," undated leaflet (approximately 1912), document number 93200196, held in the collection of the Edo-Tokyo Museum.

55. Calder, *Financing the American Dream,* p. 166.

56. Ibid., pp. 74, 98–101.

57. Tokyo shōkō kaigisho, ed., *Geppu hanbai seido,* p. 227.

58. Fukushima Hachirō, "Geppu, wappu, kurejitto: Sōkan 200 gō ni yosete," *Gekkan kurejitto* no. 200 (1973): p. 20.

59. The monthly circulation of *Fujokai* had reached nearly two hundred thousand copies by the early 1920s. *Fujokai* 27, no. 1 (January 1923), "A Ten Year History," offers circulation figures.

60. Imura Nobuko, "Kodomo no nichijō fuku wo zenbu yōfuku ni," *Fujokai* 21, no. 1 (January 1920): pp. 213–15.

61. E. R. A. Seligman, *The Economics of Installment Selling,* vol. 1 (New York: Harper and Brothers, 1927), p. 267.

62. Tokyo shōkō kaigisho, ed., *Geppu hanbai seido,* pp. 1–2.

63. Carstensen, *American Enterprise in Foreign Markets,* pp. 62–64.

64. Ibid., p. 64

65. Godley, "Selling the Sewing Machine," p. 290.

66. Andrew Godley, "Consumer Durables and Westernization in the Middle East: The Diffusion of Singer Sewing Machines in the Ottoman Region, 1880–1930," paper presented at Eighth Mediterranean Social and Political Research Meeting, Florence, March 2007, pp. 5–13. Comparative per capita income is from Angus Maddison, *The World Economy: Historical Statistics* (Paris: Organization for Economic Cooperation and Development, 2003). On Ceylon, see Wickramasinghe, "The Reception of the Singer Sewing Machine in Colonial Ceylon/Sri Lanka," p. 13.

67. Imura, "Kodomo no nichijō fuku wo zenbu yōfuku ni," pp. 213–15.

68. Gakuji kankei kenmei mokuroku, "Shingā saihō in, Monbusho kaji ka-

gaku tenrankai shuppin, saihō mishin ni kansuru setsumei sho," 10/30/1918, in the Tokyo Metropolitan Public Records Office.

69. "Shingā mishin katarogu," in the collection of Edo-Tokyo Museum, item no. 91222542, undated.

70. "Jōhin na mishin shishū: Fujin ni susumetaki fukugyō," *Fujin sekai* 14, no. 3 (February 1919): pp. 129–31.

3. SELLING AND CONSUMING MODERN LIFE

1. Hata Toshiyuki, "Shingā seizō kaisha ni kan suru hōkoku," p. 2; reprinted in Matsumura Satoshi, *Kaigai jitsugyō kenshūsei hōkoku, Nōshōmushō shōkō-kyoku rinji hōkoku*, vol. 10 (Tokyo: Yumani shobō, 2002), pp. 327–28.

2. *Fujokai* 1, no. 1 (1910), ad in the frontmatter. This was one of three popular magazines of the time whose titles translate as *Women's World*. The others were *Fujin sekai* and *Fujinkai*.

3. Singer sales leaflet, undated, estimated 1910–15, in collection of the Nature and Science Museum of the Tokyo University of Agriculture and Technology.

4. *Fujin sekai* 8, no. 5 (1913), ad in frontmatter.

5. *Fujin sekai* 8, no. 7 (1913), ad in frontmatter.

6. After 1913, the company placed no ads in the many magazines or newspapers I have surveyed.

7. "Katei oyobi shokugyō yō shingā mishin mokuroku" (Edo-Tokyo Museum, item no. 93200196). This undated brochure is stamped with the mark of the "Nagano local store." The text notes the temporary closing of the Singer Sewing Academy in "September 1910" and its reopening "at the end of 1911," suggesting that it was published in 1912. The figure of 2 million Singer machines sold globally also places the document at 1912–1914.

8. "Shingā mishin katarogu" (Edo-Tokyo Museum, item no. 91222542, tentatively dated 1922).

9. Singer sewing machine contract, 11/27/1925, signed by Yoshikawa Nomiko (Edo-Tokyo Museum, item no. 90364917).

10. For comparable duality in France, see Judith Coffin, "Credit, Consumption, and Images of Women's Desires: Selling the Sewing Machine in Late Nineteenth-Century France," *French Historical Studies* 18 (Spring 1994): pp. 749–83.

11. I have found this same tableau set in Burma and in Java, in Singer advertising postcards. Unlike the Japanese illustration, these may have been disseminated only to an American audience.

12. On the new fathers of early twentieth-century Japan, see Harald Feuss, "Men's Place in the Women's Kingdom: New Images of Fatherhood in Taishō Japan," in *Public Spheres, Private Lives: Essays in Honor of Albert M. Craig*, ed. Gail Lee Bernstein, Andrew Gordon, and Kate Wildman Nakai (Cambridge, MA: Harvard Asia Center Monographs, 2005), pp. 259–92. On the modernity of the notion of a "family circle" (*ikka danran*), see Jordan Sand, *House and Home in Modern Japan: Architecture, Domestic Space, and Bourgeois Culture, 1880–1930* (Cambridge, MA: Harvard Asia Center Monographs, 2003), pp. 29–39.

13. *Fujinkai* 1, no. 1 (1909).

14. Barbara Sato, in *The New Japanese Woman* (Durham, NC: Duke Uni-

versity Press, 2003, pp. 78–113), offers a helpful general survey of women's magazines and their readers.

15. "Nihon fuku ni wa doko ni kakushi fukuro o tsukeru ga yoki ka," *Fujin sekai* 3, no. 13 (1908): pp. 100–104. First prize was five yen.

16. "Wagaya ni okeru haibutsu riyō no keiken," *Fujokai* 22, no. 1 (1920): p. 59.

17. These stories fall into Sato's category of "practical articles," which also includes instructional pieces directed at readers. They are distinct from what she describes as "confessional" articles, which offered titillating accounts of the travails of love and marriage quite removed from the experience of most readers (Sato, *New Japanese Woman*, pp. 100–103, 105–8). See Miriam Silverberg, *Erotic Grotesque Nonsense* (Berkeley and Los Angeles: University of California Press, 2006), p. 6, on the impossibility of documenting the "actual response" of these readers.

18. Osaka shi, shakai bu chōsa ka, ed., *Yoka seikatsu no kenkyū: Rōdō chōsa hōkoku 19* (Kyoto: Kōbundō, 1923), pp. 238–316.

19. Nihon toshokan kyōkai, ed., *Shokugyō fujin dokusho keikō chōsa* (Tokyo: Nihon toshokan kyōkai, 1935), pp. 5–11.

20. Christopher Breward, "Patterns of Respectability: Publishing, Home Sewing and the Dynamics of Class and Gender, 1870–1914," similarly stresses the "variety of messages" carried in Britain by dressmaking patterns and the magazines that distributed them (p. 23).

21. "Icchaku san en de dekita shōjo yō doresu," *Fujokai* 34, no. 4 (1926): pp. 295–97.

22. Imura Nobuko, "Kodomo no nichijō fuku o zenbu yōfuku ni," *Fujokai* 21, no. 1 (1920): p. 113.

23. "Wagaya ni okeru haibutsu riyō no keiken," *Fujokai* 22, no. 1 (1920): p. 59.

24. "Ifuku, futon, shokuryōhin ni oite no kufū," *Fujokai* 22, no. 4 (1920): p. 94.

25. "Gesshū hyakuen no ginkōin no kakei," *Fujokai* 42, no. 1 (1930): pp. 163–65.

26. "Hasukubana mishin," *Fujokai* 32, no. 1 (1925): p. 261.

27. "Hoomu mishin," *Fujokai* 41, no. 1 (1930): p. 309 (and other issues).

28. Matsunami Tadayuki, *Sugu yaki ni tatsu geppu hanbai hō* (Tokyo: Banrikaku shobō, 1930), pp. 148, 160.

29. Sato notes the "conservative" outlook of magazine editors, while recognizing its potential to unintentionally provoke "small upheavals" in women's lives (*New Japanese Woman*, p. 101).

30. Watanabe Shigeru, "Mishin kawan ka, saiken kawan ka," *Fujin kurabu* 7, no. 5 (1926): pp. 257–59.

31. Ōnuma Jun, *Bunka fukusō gakuin yonjū nen no ayumi* (Tokyo: Bunka fukusō gakuin: 1963), p. 4.

32. Watanabe, "Mishin kawan ka, saiken kawan ka," pp. 257–59.

33. Silverberg, *Erotic Grotesque Nonsense,* p. 144.

34. On café waitresses, see ibid., ch. 2. On the anxious response, see Sheldon Garon, *Molding Japanese Minds: The State in Everyday Life* (Princeton, NJ: Princeton University Press, 1997), ch. 3.

35. Yoshimoto Yōko, an independent scholar affiliated with the *Shōwa no kurashi hakubutsu kan* (Showa Era Daily Life Museum) has analyzed sixty-four articles in this "true story" genre that were published in *Shufu no tomo* between 1932 and 1955 and focused specifically on "Western-clothing homeworkers," broadly defined to include women who opened their own dress shops. I discuss these in chapter 5. See Yoshimoto Yōko, "Onna no jiritsu o sasaeta yōsai," in *Yōsai no jidai: Nihon no ifuku kakumei,* ed. Koizumi Kazuko (Tokyo: OM shuppan, 2004), pp. 49–50

36. Fujiko, "Gesshū 45 en no mishin nui naishoku," *Shufu no tomo,* October 1918, pp. 125–26.

37. Ibid.

38. Forming such a group rather than a business might have helped shield earnings from the tax authorities.

39. Tenkyō Mitsuko, "Kodomo fuku saihō de sen-en no chokin," *Fujin kurabu* 6, no. 10 (1925): pp. 216–17.

40. Miyoshi Hisae, "Kodomo fuku no saihō mise o hajimete," *Fujokai* 45, no. 2 (1932): pp. 282–84.

41. Nakai Yaeko, "Gunju hin no mishin naishoku wo shite," *Fujokai* 50, no. 1 (1934): pp. 139–41.

42. Yokoyama Gennosuke, *Nihon no kasō shakai* (Tokyo: Iwanami shoten, 1949), a reprint of an 1899 publication, lists "the extraordinarily numerous varieties of homework for poor households, such as wrapping cigarettes, pasting covers on matchboxes, preparing lamp shades, making brushes, sewing *tabi,* sewing sandal thongs, padding sandal thongs, preparing envelopes, sewing handkerchiefs or knitting" (pp. 44–45). See also Matsubara Iwagorō, *Saiankoku no Tokyo* (1893; repr., Tokyo: Iwanami shoten, 1988), p. 36 on Poverty University; pp. 35, 158 on homework.

43. Tokyo shi, shakai kyoku, *Naishoku ni kansuru chōsa* (Tokyo: Tokyo shi, shakai kyoku, 1926), p. 17. The survey was conducted in 1925 and published the next year. The Social Bureau also published studies of homework in 1921 and 1935.

44. Ibid., pp. 9–10.

45. Ibid., p. 8.

46. Ibid., pp. 17–18.

47. Tōbu teishin kyoku, *Naishoku shōrei ni kan suru jikō* (Tokyo: Teishin kyoku, 1915), p. 43.

48. Ibid., p. 1.

49. Ibid., pp. 2–6.

50. Ibid., pp. 7–10.

51. The National Diet Library collection holds eleven "how to work at home" books published between 1918 and 1921. Some of these were originally serialized in newspapers, and certainly not all of these ephemeral popular tracts were collected.

52. Takahashi Keiji, *Fujin katei naishoku* (Tokyo: Seikadō, 1919), p. 9.

53. Ibid., pp. 21–23.

54. Mitani Hiroshi, "A Protonation-State and Its 'Unforgettable Other,'" in

New Directions in the Study of Meiji Japan, ed. Helen Hardacre with Adam L. Kern (Leiden and New York: Brill, 1997), pp. 301–5.

55. Hata Rimuko, *Mishin saihō hitori manabi,* 3rd ed. (Tokyo: Hata shoten shuppan bu, 1933), p. 6.

56. Souda Satoru, *Saihō mishin shiyō hō zensho* (Tokyo: Souda shuppan bu, 1930), p. 4 of main text for Souda's claim; p. 2 of frontmatter for newspaper endorsements.

57. Ushigome Chie, "Saihō kyōiku kaisō 50 nen," in *Gakuen* (Shōwa joshi daigaku), no. 245 (July 1960): p. 23.

58. Ibid., p. 24.

59. Kuwata Naoko, "Shimin yōsai fukyū katei ni okeru saihōka no tenkō to jirema: Narita Jun no yōsai kyōiku ron o chūshin ni," *Kyōiku gaku kenkyū* 65, no. 2 (1998): pp. 122–23.

60. Ibid., pp. 123, 128.

61. Sakai Nobuko, "Kinoshita Takeji to saihō gakushū ron," *Katei kagaku,* no. 76 (1978): pp. 45–48, 50–52.

62. Tetsuko Higuchi, "Waga kuni ni okeru hifuku kyōiku hatten no yōsō," *Kaseigaku zasshi* 30, no. 4 (1979): p. 383.

63. On tatami, see Sarah Teasley, "The National Geographics of Design: The Rhetoric of *Tatami* in 1920s and 1930s Japanese Interiors," in *De-Placing Difference: Architecture, Culture, and Imaginative Geography,* ed. Samar Akashi (Adelaide, Australia: Center for Asian and Middle Eastern Architecture, 2002), pp. 267–76.

64. On these campaigns, see Garon, *Molding Japanese Minds,* pp. 10–15. On housing reform, see Sand, *House and Home,* pp. 16–19, 181–202.

65. Ozaki Yoshitarō and Ozaki Gen, *Kore kara no saihō: Keizai kaizen* (Tokyo: Nihon fukusō kaizen kai shuppan bu, 1921).

66. Inoue Masato, *Yōfuku to Nihonjin: Kokumin fuku to iu moodo* (Tokyo: Kōsaidō shuppan, 2001), pp. 143–45, 221.

67. Hata, *Mishin saihō hitori manabi,* pp. 3–4 (preface to 2nd ed.).

68. Hata Rimuko, "Shingā saihō jogakuin no sakkon," *Tokyo Asahi shinbun,* 9/11/1907, p. 6.

69. Works by Tanizaki that evoke this spirit include *Some Prefer Nettles,* first published in Japanese in 1928; and *In Praise of Shadows,* first published in Japanese in 1933.

70. Shiohara Chiyoko, "Mishin yori hayaku nueru Shiohara-shiki no shin saihō," *Fujin sekai,* June 1921, pp. 137–38.

71. Andō Yoshinori, "Kindai Nihon ni okeru shintai no 'seijigaku' no tame ni," *Kyōiku shakaigaku kenkyū,* no. 60 (1997): pp. 100, 105–6.

72. Ibid., p. 107.

73. Nishimura Ayako and Fukuda Sumiko, "Kōtō jogakuseitō no fukusō hensen ni tsuite no hito kōsatsu," *Nihon no kyōiku shigaku,* October 1989, p. 54.

74. Ibid., p. 54.

75. Ibid., p. 56.

76. Andō, "Kindai Nihon ni okeru shintai," p. 108

77. For a fascinating account of these negotiations at eight different girls'

schools, see Hasuike Yoshiharu, "Kindai kyōikushijo ni mita jogakusei no fukusō no hensen (3)," *Kobe gakuin joshi tanki daigaku kiyō*, no. 15 (1982): pp. 71–78.

78. See Nakayama Chiyo, *Nihon fujin yōsō shi* (Tokyo: Yoshikawa kōbunkan, 1987), pp. 397–98.

79. *Fujin kurabu* 4, no. 3 (1923): pp. 244–46.

80. "Fukusō kairyō zehi kono sai dankō shitai," *Yomiuri shinbun*, 10/2/1923, p. 3.

81. "Shokugyō fujin no fukusō mondai no hihan," *Fujokai* 29, no. 4 (1924): pp. 45, 50.

82. Ibid., p. 44.

83. Ibid., pp. 46–47.

84. This view anticipates the widespread idea that women in post–World War II workplaces had an important role to play as "office flowers" who provided a pleasant work environment for men.

85. "Suso no midare o ki ni shite muzamuza shinda onna ten'in," *Tokyo Asahi shinbun*, 12/23/1932, p. 5.

86. Ono Tamae, "Ifuku no hanashi," *Yūai fujin*, no. 10 (October 1917): pp. 22–25.

87. Ishihara Osamu, *Rōdō eisei* (Tokyo: Sugiyama shoten, 1922), p. 199.

88. See various issues of *Rōdō* and the special issue of *Asahi gurafu*, 11/6/1929.

89. Tōyō bōseki kabushiki kaisha shashi hensan shitsu, ed., *Hyaku nen shi: Tōyōbō*, vol. 1 (Tokyo: Tōyō bōseki kabushiki kaisha, 1986), p. 280.

90. Kyoto-fu kōjō eisei kai, ed., "Onna kōshu fukusō no kenkyū," *Kōjō eisei shiryō*, no. 34 (1934).

91. "Yōsai no sanpi," *Tokyo Asahi shinbun*, 3/2/1925, p. 2, for Hirata; 3/3/1925, p. 2, for Uchida.

92. Ibid., 3/5/1925, p. 2.

93. Ibid., 3/8/1925, p. 2.

94. On *appappa*, see Nakayama, *Nihon fujin yōsō shi*, pp. 380–84; and Inoue, *Yōfuku to Nihonjin*, pp. 148–52.

4. RESISTING YANKEE CAPITALISM

1. "Grew Asks Guard for Singer Branch," *New York Times*, 1/19/1933, p. 8.

2. For Singer's statement of 11/12/1932, see Gaimushō kiroku, "Honpō ni okeru rōdō sōgi kankei zakken: 'Shingā mishin' kaisha kankei," pp. 255–56 (National Archives of Japan, Japan Center for Asian Historical Records, Tokyo). This source is hereafter cited as Gaimushō kiroku. For the employee statement, see Gaimushō kiroku, 11/15/1932, pp. 322–23.

3. Tokyo Police Chief to Home Minister, 3/19/1925, in "Kyū-kyōchōkai shiryō: Rōdō sōgi, 1925 (9)," held at the Ohara Institute for Social Research.

4. Tokyo Police Chief to Home Minister, 3/30/1925, in ibid.

5. Kamei Nobuyuki, Kyōchōkai Osaka Office director, report to Kyōchōkai General Manager Soeda Juichi, 12/17/1925, in ibid.

6. These claims are made in two leaflets dated 12/16/1925, held in "Kyūkyō-

chōkai shiryō chōkai shiryō: Sono hoka no rōdō kumiai, Shingā kaisha seido kakusei dōmei" (Call number V-16 at the Ohara Institute of Social Research).

7. A copy of this lawsuit is included in ibid.

8. On the return to and abandoning of the gold standard, see Mark Metzler, *Lever of Empire: The International Gold Standard and the Crisis of Liberalism in Prewar Japan* (Berkeley and Los Angeles: University of California Press, 2006), chs. 10–12. On gold outflows, see p. 238 and appendix A.3.

9. On price trends, see Yamamoto Tōsaku, "Zen Nihon shingā mishin kaisha shain daihyō," in "Yōkyūsho," November 1932, in Gaimushō kiroku, pp. 426–28; and Shingā sōgi dan chūō honbu, "Shingā kaisha sōgi no keika to sono shinsō," 12/15/1932, in Gaimushō kiroku, p. 591.

10. On McLeary's visit and summer activities, see Imazu Kikumatsu, "Shingā mishin no sōgi, 1," in *Rōdō kenkyū*, no. 47 (November 1951): pp. 34–35. On the pay cut, see also "Uōkā shi no seimeisho o hanku su," *Shakai undō tsūshin,* 9/6/1932.

11. "Moto kyoryūchi no kaisha ni mukai demo," *Shakai undō tsūshin,* 9/8/1932; and "Shingā mishin sōgi kaiketsu," *Shakai undō tsūshin,* 9/13/1932.

12. Tokyo police report, 1/21/1933, Gaimushō kiroku, p. 759.

13. Kanagawa police report, 10/3/1932, Gaimushō kiroku, pp. 8–9.

14. Kanagawa police report, 10/22/1932, Gaimushō kiroku, pp. 17–18.

15. "Yōkyūsho," November 1932, Gaimushō kiroku, p. 429.

16. Kyōchōkai, ed., *Saikin no shakai undō* (Tokyo: Kyōchōkai, 1927), p. 776.

17. "Yōkyūsho," November 1932, Gaimushō kiroku, p. 429.

18. Ibid., pp. 445–62.

19. Ibid., p. 457. A *mon* was a Tokugawa-era coin, comparable to a *sen* or penny.

20. On the breakdown of negotiations on the October 24, see "Kyō tsui ni sōhigyō," *Yomiuri shinbun,* 1/26/1932, evening ed., p. 2; Tokyo Police Chief report, 10/16/1932, Gaimushō kiroku, p. 36.

21. "Kyō tsui ni sōhigyō," *Yomiuri shinbun,* 10/26/1932, evening ed., p. 2.

22. "Shingā mishin sōgi kakudai," *Tokyo Asahi shinbun,* 10/25/1932, evening ed., p. 2.

23. "Kōmyō na jikyūsen," *Yomiuri shinbun,* 10/27/1932, evening ed., p. 2; *Tokyo Asahi shinbun* (10/25/1932, evening ed., p. 2) claimed that forty-five Tokyo shops with six hundred employees supported the strike.

24. Kanagawa Governor report, 12/5/1932, Gaimushō kiroku, p. 500.

25. Kanagawa Governor report, 11/18/1932, Gaimushō kiroku, p. 320.

26. Tokyo police report, 11/25/1932, Gaimushō kiroku p. 381.

27. Osaka Governor report, 12/18/1932, Gaimushō kiroku, p. 672.

28. On the provincial situation, see Gaimushō kiroku, pp. 49, 54, 74, 76; and reports from Hokkaido Governor to Home Minister of 10/26, 11/7, 11/9, 11/11, 11/14, 11/25, 12/1, Gaimushō kiroku, pp. 479–87.

29. Report from Tokyo police to Home Minister, 11/5/1932, Gaimushō kiroku, pp. 137–49.

30. Report from Kanagawa Governor to Home Minister, 11/6/1932, Gaimushō kiroku, p. 151.

31. Report of Tokyo police to Home Minister, 12/5/1932, Gaimushō kiroku, pp. 513–14.

32. Report of Osaka Governor to Home Minister, 12/28/1932, Gaimushō kiroku, p. 678–79; undated Foreign Ministry report, Gaimushō kiroku, p. 732.

33. "Singer Office Says Police Knew Mob Planned to Attack," *Japan Advertiser,* 1/20/1933, clipping in Gaimushō kiroku, p. 817.

34. "Kaisha gawa, niwaka ni yō," *Chūgai shinbun,* 10/27/1932, clipping in Gaimushō kiroku, p. 60.

35. "Singer Office Says Police Knew Mob Planned to Attack."

36. Tokyo Police Chief report to Home Minister, 11/18/1932, Gaimushō kiroku, p. 308.

37. Letter from Singer to store managers, 11/28/1932, reproduced in Tokyo Police Chief report to Home Minister, 12/5/1932, Gaimushō kiroku, pp. 517–21.

38. Kanagawa Governor report to Home Minister, 1/12/1933, Gaimushō kiroku, p. 707.

39. Tokyo Police Chief report to Home Minister, 1/19/1933, Gaimushō kiroku, p. 718–20.

40. "Sōgi ga iya ni nari, hito shibai utasu," *Yomiuri shinbun,* 1/14/1933, p. 7.

41. Kanagawa Governor report to Home Minister, 11/30/1932, Gaimushō kiroku, p. 414, on hiring of one group in Kobe; and 12/13/1932, p. 549, on hiring of another in Yokohama. On violence over the course of the dispute, see Kanagawa Governor report to Home Minister, 11/18/1932, Gaimushō kiroku, p. 300; Kanagawa Governor report to Home Minister, 12/6/1932, ibid., p. 523; Tokyo police report to Home Minister, 12/7/1932, ibid., p. 528; Kanagawa Governor report, 12/8/1932, ibid., pp. 535–38; and Osaka Governor report to Home Minister, 12/13/1932, ibid., p. 572.

42. Kanagawa police report, 1/21/1933, Gaimushō kiroku, pp. 778–87.

43. "Strikers Wreck Singer Office," *Japan Advertiser,* 1/21/1933, clipping in Gaimushō kiroku, pp. 818–19.

44. Kanagawa Governor report to Home Minister, 1/28/1933, Gaimushō kiroku, pp. 883–94; Kanagawa Governor report to Home Minister, 2/3/1933, ibid., p. 901.

45. Report from Japanese ambassador to the United States to Foreign Minister Uchida, 1/18/1933, Gaimushō kiroku, p. 713; *Japan Advertiser,* 1/23/1933, on Grew requests to Foreign Ministry.

46. For details of the settlement, see *Tokyo Asahi shinbun,* 2/9/1933, p. 7, clipping in Gaimushō kiroku, p. 912; Tokyo Police report to Home Minister, 2/13/1933, ibid., pp. 913–15; and Kanagawa Governor report to Home Minister, 2/18/1933, ibid., pp. 916–32.

47. On the fate of Tokyo employees, see Tokyo police report to Home Minister, 3/28/1933, Gaimushō kiroku, pp. 961–66. For Osaka, see Osaka Governor report to Home Minister, 3/16/1933, ibid., pp. 955–58.

48. "Demo de oshikake, misankasha o kanyū," *Shakai undō tsūshin,* 11/22/1932.

49. Report of Tokyo Police Chief to Home Minister, 11/5/1932, Gaimushō kiroku, 139.

50. Report of Osaka Governor to Home Minister, 10/28/1932, Gaimushō kiroku, 75.

51. Report of Saga Prefecture Governor to Home Minister, 12/24/1932, transcribes leaflet sent to shops in the prefecture, Gaimushō kiroku, pp. 619–20.

52. "Ni jikan amari ni watari, gekiron o majie, ketsuretsu su," *Shakai undō tsūshin,* 10/26/1932.

53. Report of Kanagawa Governor to Home Minister, 10/27/1932, Gaimushō kiroku, pp. 54–57. The expression translated here as "gone foreign" is *tōka.* The character *tō* refers to Tang dynasty China and to China more generally. It also carries a pejorative sense of foreignness. A more literal translation would be "Sinified" or "Tang-ified."

54. *Dong-A Ilbo,* 9/21/1932.

55. Report of Saga Prefecture Governor to Home Minister, 12/24/1932, Gaimushō kiroku, p. 619; "Yōkyūsho," November 1932, ibid., p. 429; "Sōgi no keika to sono shinjitsu," ibid., p. 589.

56. Andrew Gordon, "Business and the Corporate State: The Business Lobby and Bureaucrats on Labor, 1911–1941," in *Managing Industrial Enterprise,* William Wray, ed. (Cambridge, MA: Harvard Council on East Asian Studies Monographs, 1989), pp. 67–68.

57. For extended discussion of this topic, see Andrew Gordon, *The Evolution of Labor Relations in Japan: Heavy Industry, 1853–1955* (Cambridge, MA: Harvard Council on East Asian Studies Monographs, 1985), ch. 4.

58. "Yōkyūsho," November 1932, Gaimushō kiroku, p. 457.

59. "Sōgi no keika to sono shinjitsu," Gaimushō kiroku, p. 589.

60. "Statement" submitted to "T. Taketomi, in charge of Foreign Trade, Foreign Office, Kasumigaseki, Tokyo," 11/12/1932, Gaimushō kiroku, pp. 251, 255.

61. Robert Bruce Davies, *Peacefully Working to Conquer the World: Singer Sewing Machines in Foreign Markets, 1854–1920* (New York: Arno Press, 1976), p. 114.

62. Andrew Godley, "Selling the Sewing Machine Around the World: Singer's International Marketing Strategies, 1850–1920," *Enterprise and Society* 7, no. 2 (2006): pp. 282–83, 286–87.

63. Mona Domosh, *American Commodities in an Age of Empire* (New York: Routledge, 2006), pp. 44–45; Davies, *Peacefully Working to Conquer the World,* p. 185.

64. Fred V. Carstensen, *American Enterprise in Foreign Markets* (Chapel Hill: University of North Carolina Press, 1984), pp. 64–65.

65. Ibid., p. 81.

66. "Statement," 11/12/1932, Gaimushō kiroku, p. 254, on the refusal to talk to "group" representatives.

67. Shimazono Satoshi, *Dotō o koete* (Tokyo: privately published, 1960); pp. 159–61 recount these two episodes.

68. Osaka Governor report to Home Minister, 2/3/1933, Gaimushō kiroku, pp. 906–10.

69. On Mitsubishi and Brothers, see Kuwahara Tetsuya, "Shoki takokuseki kigyō no tainichi tōshi to minzoku kigyō," *Kokumin keizai zasshi* 185, no. 5

(2002): p. 52; and Janome mishin shashi hensan iinkai, ed., *Janome mishin sōgyō 50 nen shi* (Tokyo: Tokiwa shoin, 1971), p. 212.

70. Kuwahara, "Shoki takokuseki kigyō," p. 52.

71. "Ki ni nojite, kokusan hin shinshutsu," *Shakai undō tsūshin,* 11/22/1932.

72. Report of Osaka Governor to Home Minister, 12/28/1932, Gaimushō kiroku, pp. 674, 677; Janome mishin shashi hensan iinkai, ed., *Janome mishin sōgyō 50 nen shi,* p. 211.

73. Janome mishin shashi hensan iinkai, ed., *Janome mishin sōgyō 50 nen shi,* p. 212.

74. Shimazono, *Dotō o koete,* pp. 163–75. See also Burazaa kōgyō kabushiki kaisha, *Burazaa no ayumi* (Tokyo: Dayamondo sha, 1971), pp. 49–52.

75. Janome mishin shashi hensan iinkai, ed., *Janome mishin sōgyō 50 nen shi,* p. 243.

76. Ibid., pp. 244–55.

77. Ibid., p. 246 (excerpts from the manual, which unfortunately does not survive).

78. Ibid., p. 277.

79. On the layaway plan, see ibid., pp. 197–200. For the Shimada quotation, see p. 239.

80. Lendol Calder, *Financing the American Dream,* pp. 195–99.

81. On rural credit societies in the 1930s, see John Embree, *Suyemura: A Japanese Village* (Chicago: University of Chicago Press, 1939), pp. 138–47. On Ose's view of this link, see Janome mishin shashi hensan iinkai, ed., *Janome mishin sōgyō 50 nen shi,* pp. 197, 199–200.

82. On subsidies and awards to Janome, see Janome mishin shashi hensan iinkai, ed., *Janome mishin sōgyō 50 nen shi,* pp. 191–92.

83. "Kokusan shōrei no hinmoku kimaru: 118 shū ni," *Yomiuri shinbun,* 3/10/1928, p. 2; "Yūryō kokusan hin chikaku happyō, Shōkō shō sentei no 312 hinshū," *Yomiuri shinbun,* 4/21/1930, p. 2.

84. Mark Mason, *American Multinationals and Japan: The Political Economy of Japanese Capital Controls, 1899–1980* (Cambridge, MA: Harvard East Asia Monographs, 1992); p. 57 offers one indication of Singer's weak position by 1938: it was even denied an import license to replace parts damaged in a flood in Kobe.

85. On GM, Ford, and other American companies in 1930s Japan, see ibid., pp. 60–99.

86. Walter Friedman, *Birth of a Salesman* (Cambridge, MA: Harvard University Press, 2004), p. 269.

5. WAR MACHINES AT HOME

1. Minami Hiroshi and Shakai shinri kenkyūjo, eds., *Shōwa bunka, 1925–1945* (Tokyo: Keisō shobō, 1987), p. ii.

2. See Minami Hiroshi, ed., *Kindai shōmin seikatsu shi,* vol. 5 (Tokyo: Sanichi shobō, 1986), pp. 544–45, on the militarization of dress. For insightful discussion of Minami's views, see Inoue Masato, *Yōfuku to Nihonjin: Kokumin fuku to iu moodo* (Tokyo: Kōsaidō shuppan, 2001), p. 28.

3. Louise Young, *Japan's Total Empire: Manchuria and the Culture of Wartime Imperialism* (Berkeley and Los Angeles: University of California Press, 1998), p. 435; Miriam Silverberg, *Erotic Grotesque Nonsense: The Mass Culture of Japanese Modern Times* (Berkeley and Los Angeles: University of California Press, 2006), p. 5. See also, Janice Mimura, *Planning for Empire: Reform Bureaucrats and the Japanese Wartime State, 1931–1945* (Ithaca, NY: Cornell University Press, 2011).

4. Nihon hōsō kyōkai, ed., *Rajio nenkan* (Tokyo: NHK shuppan, 1940), p. 270.

5. Nakayama Chiyo, *Nihon fujin yōsō shi* (Tokyo: Yoshikawa kōbunkan, 1987), p. 416.

6. Young, *Japan's Total Empire*, pp. 57–68.

7. Cited in Nakayama, *Nihon fujin yōsō shi*, pp. 397–98.

8. Kon Wajirō, "Wa-yōsō hiritsu suitei," *Tokyo shūhō*, 3/5/1933, in *Kon Wajirō shū dai 8 kan: Fukusō shi* (Tokyo: Domesu shuppan, 1976), p. 167.

9. "Sutairu bukku no fukkyū, *Sharuman*, September 1935, in ibid., p. 168.

10. Kon Wajirō, "Zenkoku 19 toshi josei fukusō chōsa hōkoku," *Fujin no tomo*, June 1937, pp. 89–113.

11. Nakayama, *Nihon fujin yōsō shi*, p. 414.

12. On the difference between schools opened by tailors and by women, see Yoshimoto Yoko, "Yōsai gakkō," in *Yōsai no jidai: Nihon no ifuku kakumei*, ed. Koizumi Kazuko (Tokyo: OM Shuppan, 2004), p. 28. Yoshimoto is an independent scholar affiliated with the Showa Era Daily Life Museum. On the more general expansion of Western dressmaking schools, see Nakayama, *Nihon fujin yōsō shi*, pp. 417–18.

13. "Shokugyō o nozomu fujin no gakkō annai," *Fujin kurabu*, January 1934, pp. 420–25.

14. Ibid.

15. Inoue, *Yōfuku to Nihonjin*, p. 201.

16. Nakayama, *Nihon fujin yōsō shi*, pp. 423–25.

17. Nihon mishin kyōkai, ed., *Nihon mishin sangyō shi* (Tokyo: Nihon mishin kyōkai, 1961), pp. 38–39.

18. Tokyo shōkō kaigisho, *Geppu hanbai seido* (Tokyo: Tokyo shōkō kaigisho, 1929), pp. 211–12.

19. Tokyo shiyakusho kōgyō kyoku shōgyō ka, ed., *Wappu hanbai ni kan suru chōsa* (Tokyo: Tokyo shiyakusho,1935), p. 62.

20. Barbara Burman, "Made at Home By Clever Fingers: Home Dressmaking in Edwardian England," in *The Culture of Sewing: Gender, Consumption and Home Dressmaking*, ed. Barbara Burman (Oxford: Berg, 1999); pp. 33–34 identify a similar package that coalesced in Britain in the early 1900s.

21. On Shimada, see Janome mishin sha shi hensan iinkai, ed., *Janome mishin sōgyō gojū nen shi* (Tokyo: Janome mishin kōgyō kabushiki kaisha, 1971), pp. 253–54. A list of the ads can be quickly generated from the *Asahi shinbun* Shōwa senzen database and the *Yomiuri shinbun* Shōwa senzen databases 1 and 2.

22. *Tokyo Asahi shinbun*, 1/9/1936, p. 5.

23. *Tokyo Asahi shinbun*, 1/5/1938, p. 7, and 3/21/1939, p. 7.

24. These ads appeared, in the order introduced here, in *Tokyo Asahi shinbun*, 8/23/1935, p. 5; 2/23/1936, p. 4; 1/29/1938 evening ed., p. 4; 1/2/1939,

p. 10; and 12/9/1941, p. 10. Self-cultivation as a modern value associated with a Taisho-era spirit of middle-class individualism is discussed in Takeuchi Yō, *Risshin shusse to nihonjin* (Tokyo: NHK shuppan, 1996), p. 162; and Barbara Sato *The New Japanese Woman* (Durham, NC: Duke University Press, 2003), pp. 134–48.

25. *Fujokai,* April 1933, unpaginated pictorial insert.

26. *Fujin kurabu,* January 1934, pp. 420–21.

27. *Fujin kurabu,* December 1938, p. 187.

28. Nakayama, *Nihon fujin yōsō shi,* p. 419.

29. For the text analysis summarized here, see Yoshimoto Yōko, "Onna no jiritsu o sasaeta yōsai." She also looked at twenty-five stories from the first post-war decade.

30. Ibid., p. 50.

31. Nihon hōsō kyōkai, *Kokumin seikatsu jikan chōsa: Hōkyū seikatsu sha, kōjō rōmu sha, joshi kazoku hen* (Tokyo: Nihon hōsō kyōkai, 1943), pp. 90–93. See the appendix for detailed discussion.

32. See Ishikawa bunka jigyō zaidan, Ochanomizu toshokan, ed., *Shufu no tomo: Shōwa ki mokuji,* vol. 2 (Tokyo: Ochanomizu toshokan, 2009) for wartime tables of contents of *Shufu no tomo.*

33. "Katei no fukugyō o motomeru hito ni chōhō na jusanjo no shigoto to riyōhō," *Fujokai,* February 1935, pp. 234–37.

34. Ibid.

35. "Shufu-tachi no naishoku-bayari," *Tokyo Asahi shinbun,* 5/21/1935, p. 5.

36. Yoshimoto, "Onna no jiritsu wo sasaeta yōsai," pp. 50–52.

37. Ibid., pp. 54–57.

38. Homma Yuriko, "Kodomo no gakuhi o tsukutta mishin naishoku," *Shufu no tomo,* April 1935, pp. 335–38. This story is one of three printed together under the title "The Experience of Women Who Succeeded with Sewing Machine Homework." The others are Iida Kazuko, "Jogakkō no fukin de yōsōten o kaigyō," pp. 332–34; and Hata Tomoko, "Kōgai no shinkaichi de hajimeta epuron ten," pp. 338–40, both stories of women successfully opening dress shops. See also similar sentiments in the monthly advice column of *Fujokai,* August 1939, p. 363.

39. Gregory J. Kasza, *The State and the Mass Media in Japan, 1918–1945* (Berkeley and Los Angeles: University of California Press, 1988), pp. 232–42.

40. Thomas Havens, *Valley of Darkness: The Japanese People and World War II* (New York: W. W. Norton, 1998), p. 67.

41. Silverberg, *Erotic Grotesque Nonsense,* p. 1.

42. Wm. Theodore de Bary, Carol Gluck, and Arthur E. Tiedemann, eds., *Sources of Japanese Tradition: 1600 to 2000,* 2nd ed., vol. 2 (New York: Columbia University Press, 2005), pp. 968–75.

43. *Yomiuri shinbun,* 7/5/1937, p. 7. The law to conserve electric power of October 1939 only prohibited use of permanent wave machines in private homes. See "Raigetsu tsuitachi kara kinshi," *Yomiuri shinbun,* 7/7/1939, evening ed., p. 2. See also *Yomiuri shinbun,* 8/28/1943, p. 3; and 8/29/1943, p. 2.

44. "Sesō no han'ei," *Tokyo Asahi shinbun,* 11/8/1935, p. 9.

45. These articles were identified via a keyword search for *geppu,* using the *Asahi shinbun senzen shōwa deeta beesu.* Thirteen of the stories concerning tax

issues, foreign practice, and other aspects could not be categorized as positive or negative.

46. *Tokyo Asahi shinbun,* 2/11/1930, p. 4 ; 3/4/1933, p. 5; 9/18/1933, p. 5.

47. The *Yomiuri shinbun* articles were also identified via a keyword search for *geppu,* using the *Yomiuri shinbun senzen shōwa deeta beesu I (1926–1936) and II (1936–1945).* This database also identified 1,350 advertisements for products sold on installment, generally either radios, sewing machines, or cameras.

48. "Hōkyūsha ni totte wa benri na geppu hensai," *Tokyo Asahi shinbun,* 4/14/1938, p. 6.

49. Matsumiya Saburō, *Sugu kiku kōkoku* (Tokyo: Mikasa shobō, 1938), pp. 77–80.

50. "Mutanpo murishi de geppu de mo kaeseru," *Yomiuri shinbun,* 8/7/1939, p. 5.

51. "Yōfuku shitate chin 5 bun sage," *Tokyo Asahi shinbun,* 4/7/1939, p. 10.

52. "Yōfuku no geppu wa maru-kō yori 5 bun mashi: Hatsuka kara jisshi," *Yomiuri shinbun,* 7/14/1941, p. 3.

53. Nihon mishin kyōkai, ed., *Nihon mishin sangyō shi,* pp. 40–41.

54. Janome mishin sha shi hensan iinkai, ed., *Janome mishin sōgyō gojū nen shi,* pp. 286–87.

55. Nihon mishin kyōkai, ed., *Nihon mishin sangyō shi,* pp. 41–49 for this overview; Janome mishin sha shi hensan iinkai, ed., *Janome mishin sōgyō gojū nen shi,* pp. 297–300 on the Janome company efforts; p. 809 for aggregate production statistics.

56. Janome mishin sha shi hensan iinkai, ed., *Janome mishin sōgyō gojū nen shi,* pp. 300–301.

57. Taylor Atkins, *Blue Japan* (Durham, NC: Duke University Press, 2001), pp. 127–63.

58. Kon Wajirō, "Shōwa 12 nen no wayōsō," *Seikatsu to shumi,* May 1937; reprinted in *Kon Wajirō shū,* vol. 8, p. 171–72.

59. Kon Wajirō, "Shōwa 13 nen no wayōsō," *Osaka mitsukoshi,* February 1938; reprinted in *Kon Wajirō shū,* vol. 8, p. 173–74.

60. Ibid., p. 176.

61. Sakai, cited in Kuwata Naoko, "Shimin yōsai fukyū katei ni okeru saihōka no tenkō to jirema: Narita Jun no yōsai kyōiku ron o chūshin ni," *Kyōiku gaku kenkyū,* June 1998, p. 6.

62. Narita Jun, "Yōsai jugyō o kaerimite" (1933), cited in Kuwata, "Shimin yōsai," p. 6.

63. Narita Jun, "Saihō no shōrai ha dō kangaeru ka" (1936), cited in Kuwata, "Shimin yōsai," p. 6.

64. "Kaji to saihō ga gōryū, atarashii hifuku ka no seikaku," *Yomiuri shinbun,* 9/17/1943, p. 4. See also related articles published on 9/16/1943, p. 3; 9/17/1943, p. 4

65. Kuwata, "Shimin yōsai," p. 7; Inoue, *Yōfuku to Nihonjin,* pp. 38–39.

66. Inoue, *Yōfuku to Nihonjin,* p. 7.

67. Inoue, *Yōfuku to Nihonjin,* p. 88; Nakayama, *Nihon fujin yōsō shi,* p. 437.

68. Kon Wajirō, "Seifuku undō sengen, *Seifuku undō,* 1/1/1939; reprinted in *Kon Wajirō shū,* vol. 8, p. 187.

69. Inoue, *Yōfuku to Nihonjin,* pp. 45–46, 48; Nakayama, *Nihon fujin yōsō shi,* p. 438.

70. Inoue, *Yōfuku to Nihonjin,* pp. 48–50; Nakayama, *Nihon fujin yōsō shi,* pp. 438–42.

71. Inoue, *Yōfuku to Nihonjin,* p. 44.

72. Nakayama, *Nihon fujin yōsō shi,* p. 442; Inoue, *Yōfuku to Nihonjin,* pp. 72–73.

73. Inoue, *Yōfuku to Nihonjin,* p. 55. The order for Standard Dress came in a "jikan kaigi ryōkai jiko."

74. Kon Wajirō, "Kokumin fuku zehi ron" *Fujin Asahi,* February 1939; reprinted in *Kon Wajirō shū,* vol. 8, p. 192.

75. Kon Wajirō, "Fujin kokuminfuku no mondai," *Hokkaidō taimusu,* 1/1/1941; reprinted in *Kon Wajirō shū,* vol. 8, p. 195.

76. Inoue, *Yōfuku to Nihonjin,* p. 55.

77. Ibid., pp. 67–69.

78. Nakayama, *Nihon fujin yōsō shi,* pp. 444–47, for drawings and a list of these types.

79. "Ifukukai dayori,"*Ifuku kenkyū* 3, no. 4 (1943): p. 55. Cited in Inoue, *Yōfuku to Nihonjin,* p. 58 (my emphasis).

80. Nakayama, *Nihon fujin yōsō shi,* p. 449.

81. "Senji ifuku mondai no ganmoku," *Yomiuri shinbun,* 6/6/1943, p. 2.

82. Kimura Matsukichi, " 'Monpe' ni kan suru kenkyū," *Hifuku* 1, no. 3 (1930), cited in Inoue, *Yōfuku to Nihonjin,* pp. 221–22.

83. On Segawa, see Liza Crihfield Dalby, *Kimono: Fashioning Culture* (New Haven, CT: Yale University Press, 1993), pp. 144–52.

84. Inoue, *Yōfuku to Nihonjin,* p. 178.

85. "Jūgo josei ni monpe kōshū," *Yomiuri shinbun,* 8/24/1938, evening ed., p. 2.

86. "Hakikata ga warui no ka, hatarakinukui monpe," *Yomiuri shinbun,* 9/7/1940, p. 4.

87. Fujii Tadatoshi, *Kokubō fujinkai: Hinomaru to kappōgi* (Tokyo: Iwanami shoten, 1985), p. 197.

88. For Saito's comment and an account of the 1941 deliberations, see Inoue, *Yōfuku to Nihonjin,* pp. 172–75.

89. Awaya, quoted in Inoue, *Yōfuku to Nihonjin,* p. 164; affirmative voices cited on pp. 215–16.

90. Inoue, *Yōfuku to Nihonjin,* pp. 190–91, quotes Iwamoto Motoko, *Hifuku yōgi: Fujin hyōjunfuku hen* (Tokyo: Hōbunkan, 1943).

91. Inoue, *Yōfuku to Nihonjin,* p. 217.

92. Helen Reynolds, " 'Your Clothes Are Materials of War': The British Government Promotion of Home Sewing During the Second World War," in Burman, *Culture of Sewing,* pp. 327–37.

93. Japan's wartime enemies in Asia also devised a hybrid form of rational modern dress for men, elites in particular, similar in appearance to Japanese Standard Dress. After the war, this became famous as the Mao suit; it traces to versions worn by Lenin, Sun Yat-sen, Stalin, and Chiang Kai-Shek. The Song sisters, for their part, improvised the cheongsam, a dress that arguably projected "modernity" and "Chineseness" more successfully than women's reform dress in Japan

signaled "modernity" and "Japaneseness." See Verity Wilson, "Dressing for Leadership in China: Wives and Husbands in an Age of Revolutions," pp. 238–58.

94. Anne Hollander, *Sex and Suits: The Evolution of Modern Dress* (New York: Alfred A. Knopf, 1994), p. 143; pp. 164–65 discuss a process stretching across two generations in which women in the West "gradually learned to wear trousers."

95. Nihon hōsō kyōkai, *Kokumin seikatsu jikan chōsa: Nōgyō setai hen,* p. 149; and *Kokumin seikatsu jikan chōsa: Hōkyū seikatsu sha, kōjō rōmusha joshi kazoku hen,* pp. 90–93.

96. Home-based sewing for brokers was counted in the large, separate category of homework (*naishoku*).

97. For one literary evocation of this scene, see Miyashita Zenji, "Hinan ressha," in *Chichi ga kataru taiheiyō senso: Moeru Nihon rettō,* vol. 3, ed. Kusuru Yoshio (Tokyo: Dōshinsha, 1969), p. 103.

98. For these trends in the United States and Britain, see Eileen Margerum, "The Needle as Magic Wand: Selling Sewing Lessons to American Girls After the Second World War," pp. 194–95; and Reynolds, " 'Your Clothes Are Materials of War,' " pp. 336–37, both in Burman, *Culture of Sewing.*

6. MECHANICAL PHOENIX

1. "Nagasaki: Genbaku o tsutaeta kisha (2)," *Asahi shinbun,* 8/10/1995, p. 3.

2. Nakayama Chiyo, *Nihon fujin yōsō shi* (Tokyo: Yoshikawa kōbunkan, 1987); p. 467 cites a Ministry of International Trade and Industry (MITI) report of 1955 based on data collected from the Japan Sewing Machine Industry Association.

3. For production data, see Nihon mishin kyōkai, ed., *Nihon mishin sangyō shi* (Tokyo: Nihon mishin kyōkai, 1961), pp. 1–5; Gilbert Burck, "Mighty Singer's New Competition," *Fortune,* February 1959, p. 106.

4. "Mishin no fukyū jōkyō to kongo no juyō ni kan suru hito kōsatsu," *Mishin kōgyō,* April 1954, p. 25.

5. Nihon mishin kyōkai, ed., *Nihon mishin sangyō shi,* p. 7.

6. Ibid., pp. 6–8.

7. Ibid., pp. 2–3.

8. Kuwahara Tetsuya, "Shoki takokuseki kigyō no tainichi tōshi to minzoku kigyō," *Kokumin keizai zasshi* 185, no. 5 (2002), pp. 54–56.

9. A pioneering work in English is Michael A. Cusumano, *The Japanese Automobile Industry: Technology and Management at Nissan and Toyota* (Cambridge, MA: Harvard Council on East Asian Studies, 1985), ch. 5.

10. Simon Partner, *Assembled in Japan: Electrical Goods and the Making of the Japanese Consumer* (Berkeley and Los Angeles: University of California Press, 1999), pp. 61–66.

11. Kuwahara, "Shoki takokuseki kigyō," p. 56, cites these memoirs.

12. *Nihon mishin taimusu* (hereafter cited as *NMT*), 3/21/1949, p. 1.

13. The list ran in four parts as "Wa ga kuni ni shiyō sarete iru mishin no shurui oyobi sono yōto," *NMT,* 3/15/1948, 4/1/1948, 4/15/1948, and 5/1/1948. For an example of the diagrams of Singer parts, see *NMT,* 1/21/1950.

14. A. K. Aurell, "Memorandum to Mr. Richard A. May," 11/5/1945, Japan War Loss Records, Singer Sewing Machine Collection, Wisconsin State Histori-

cal Society (hereafter cited simply as Japan War Loss Records). On his wartime internment, see Aurell's obituary in the *Boston Globe*, 7/20/1961, p. 30.

15. Mark Mason, *American Multinationals and Japan: The Political Economy of Japanese Capital Controls, 1899–1980* (Cambridge, MA: Harvard Council on East Asian Studies, 1992), pp. 105–11 on SCAP's resistance to American investment in occupied Japan, including Singer.

16. *NMT,* 11/1/1947, p. 3 on Lawton's return.

17. A. K. Aurell to Civil Affairs Division, 8/1/1946, Japan War Loss Records.

18. E. E. Lawton, "Memorandum," 2/19/1947, Japan War Loss Records.

19. A. K. Aurell to Sir Douglas Alexander, President, 5/2/1947, Japan War Loss Records.

20. "Excerpts from letter dated April 18th, 1947, received May 9th" (unsigned), Japan War Loss Records. This must be Aurell's letter; he was the first and only Singer employee in Japan at the time.

21. Janome mishin sha shi hensan iinkai, ed., *Janome mishin sōgyō 50 nen shi* (Tokyo: Tokiwa shoin, 1971), p. 809.

22. A. K. Aurell to Sir Douglas Alexander, President, 5/2/1947, Japan War Loss Records, for all quotations in this paragraph.

23. A. K. Aurell to W. A. Davidson, Vice President, 6/3/1947, Japan War Loss Records.

24. A. K. Aurell to E. E. Lawton, 12/30/1947, Japan War Loss Records. M.T.D. is the abbreviation for Singer's Manufacturing and Trade Division.

25. See, for example, E. E. Lawton to Major E. G. Shuck, 1/7/1948, R/G 331, entry 1639, box 4840, folder 17–2, Singer Sewing Machine Company, vol. 1, U.S. National Archives.

26. E. C. Miller Jr. to Singer Sewing Machine Company, Mr. E. E. Lawton, 12/2/1949, R/G/331, entry 1639, box 4840, folder 18–1, Singer Sewing Machine Company, vol. 2, U.S. National Archives.

27. Ministry of Finance, encl. 2, table 1, R/G 331, entry 1597, box 3672, folder 1–1, Singer Sewing Machine Company, Trademarks, Copyrights, U.S. National Archives.

28. Mason, *American Multinationals and Japan,* p. 112.

29. Unsigned letter from "Vice President" to E. E. Lawton, marked "read by D. H. Alexander," 4/28/1953, Japan War Loss Records.

30. A. K. Aurell to Singer Sewing Machine Company, New York, 6/2/1953; and F. F. Fairman to Singer Sewing Machine Company, New York, 10/5/1953, Japan War Loss Records.

31. A. K. Aurell to E. E. Lawton, 12/30/1947, Japan War Loss Records.

32. *NMT,* 2/11/1950, p. 10.

33. Kuwahara, "Shoki takokuseki kigyō," p. 59.

34. *NMT,* 6/28/1952, p. 1.

35. *NMT,* 6/21/1952, p. 2.

36. *NMT,* 6/28/1952, p. 1.

37. For sales figures, see Singer World Results, 1952, 1953, Singer Sewing Machine Collection, Wisconsin State Historical Society.

38. On rumors that installment selling was soon to begin, see *NMT,* 2/13/1953, p. 3. Singer World Reports (Singer Sewing Machine Collection, Wisconsin State

Historical Society) lists open accounts for Japan for the first time as of 12/31/1954.

39. "Shingā kyokutō shijo ni shinshutsu keikaku," *NMT,* 5/1/1953, p. 2. Follow-up stories appeared in issues of 5/8, 5/15, 5/22, 5/29, and 6/5.

40. One of the two founders of the original Pine Company, Kamematsu, had taken the Pine name with him in a dispute in 1933 with his partner. His relatively unsuccessful venture was purchased after the war by Japan Steel (Nihon seikō), formerly a small-arms producer in the Mitsui group. See Janome mishin sha shi hensan iinkai, ed., *Janome mishin sōgyō 50 nen shi,* p. 410.

41. "S-sha, P-sha e ichi oku en amari tōnyū," *NMT,* 7/16/1954, p. 1.

42. Gilbert Burck, "Mighty Singer's New Competition," *Fortune,* February 1959, p. 134.

43. "Seimeisho," *NMT,* 7/19/1954, p. 1.

44. On these laws and their impact, see Mason, *American Multinationals and Japan,* pp. 154–61; and Janome mishin sha shi hensan iinkai, ed., *Janome mishin sōgyō gojū nen shi* (Tokyo: Janome mishin kōgyō kabushiki kaisha, 1971), p. 412.

45. *Singer Manufacturing Company Annual Report, 1955* (4/26/1956), p. 3. This and subsequent Singer annual reports were obtained from the ProQuest Historical Annual Reports database.

46. "P-sha kono aki S mo seizō e," *NMT,* 4/20/1956, p. 2; Burck, "Mighty Singer's New Competition," p. 134.

47. Mason, *American Multinationals and Japan,* p. 160 on yen-based companies.

48. Shimada Takuya, "Minzoku sangyō o mamore," *NMT,* 1/6/1956, p. 3.

49. "Ni nen koshi no shingā shinshutsu, *Shūkan Tokyo* 10/6/1956, pp. 16–17.

50. *Singer Manufacturing Company Annual Report, 1956* (4/26/1957), p. 3.

51. *Singer Annual Report,* 1960, 4/4/1961, p. 15; "How Singer Sells the World Market," *Steel,* 5/9/1960, pp. 105–7.

52. On Satō's visit, see "Shingā no makikaeshi," *Zaikai,* 10/15/1959, pp. 72–74. The *Singer Annual Report,* 4/4/1961, states that partial remittance of "royalty earnings" was worked out in 1960, with assurance of fuller remittance of dividends promised for 1961 (p. 14).

53. John W. Dower, *War without Mercy* (New York: Pantheon Books, 1986), pp. 301–02.

54. "When You Buy Japanese, Double Check the Goods," *Business Week,* 12/17/1949, p. 106.

55. "Case of the Dollar Blouse," *U.S. News and World Report,* 9/14/1956, p. 67; "Men's Suits from Cheap Labor: Bad News for U.S. Industry," *U.S. News and World Report,* 10/5/1959, p. 84

56. "Fast Drive from Japan," *Time,* 8/17/1959, p. 83.

57. "When Goods from Abroad Hurt Business at Home," *U.S. News and World Report,* 3/7/1958, p. 47.

58. "Competition Returns to International Trading," *Saturday Evening Post,* 11/15/1952, p. 10.

59. Mason, *American Multinationals and Japan,* pp. 161–97 on the efforts of Coca Cola, IBM, and others.

60. Dayamondo sha, ed., *Sekai no kigyō monogatari: Shingā mishin* (Tokyo: Dayamondo sha, 1971), pp. 48–49.

61. "More Push Overseas for Singer" *Business Week,* 12/20/1958, pp. 73–4; *Singer Manufacturing Company Annual Report,* 1958 (3/31/1959), p. 2.

62. Singer had introduced zigzag machines in the 1930s but sold them only to industrial users, assuming that American women would find them too complex for home sewing; see "Regaining Its Old Time Zing," *Financial World,* 2/4/1959, pp. 6–7.

63. Interview with Ohmaki Kunichika, 6/26/2003.

64. *Singer Annual Report,* 1964 (3/19/1965), pp. 18–19.

65. *Singer Annual Report,* 1960 (4/4/1961), p. 15.

66. "Singer Company Special Report," *Forbes,* 10/15/1964, p. 24.

67. *Singer Annual Report,* 1960 (4/4/1961), p. 15.

68. "When Goods from Abroad Hurt Business at Home," p. 45.

69. Gilbert Burck, "Hardening of the Assets," *Fortune,* January 1959, pp. 148, 150.

70. On the sales strategies of electronics firms, see Partner, *Assembled in Japan,* pp. 153–68. On Matsushita, see Masahiro Shimotani, "The Formation of Distribution *Keiretsu:* The Case of Matsushita Electric," in *The Origins of Japanese Industrial Power,* ed. Etsuo Abe and Robert Fitzgerald (London: Frank Cass, 1995), pp. 54–69.

71. Janome mishin sha shi hensan iinkai, ed., *Janome mishin sōgyō gojū nen shi,* pp. 414–16.

72. Burazā kōgyō kabushiki kaisha, *Burazā no ayumi: Sekai ni idomu* (Tokyo: Dayamondo sha, 1971), pp. 118–19.

73. Dayamondo sha, ed., *Mishin: Janome* (Tokyo: Dayamondo sha, 1965), p. 135.

74. Ibid., pp. 135–37. Janome had used these exhibits in prewar days, but for general advertising rather than to accumulate prospect lists.

75. Ibid., p. 138.

76. "Sērusuman daiichi shugi," *Janome shanai hō* 1, no. 1 (1956): p. 1.

77. See *Janome shanai hō* 7, nos. 45, 46 (1962).

78. "Nōdai shiten hanbai hakusho," *Janome shanai hō* 3, no. 11 (1958): pp. 8–12.

79. "Shigoto to tomo ni ayumu," *Janome shanai hō* 7, no. 50 (1962), pp. 18–19.

80. "Mishin gyōkai to sērusuman no kimyōna kankei," *Nihon shūhō,* 4/15/1964, pp. 38–41.

81. Ibid.

82. Ibid., p. 42.

83. "Tatta san zen en no shihon de dekiru yōsai naishoku no tebiki," *Janome mōdo,* Spring 1962, pp. 6–7.

84. *NMT,* 11/11/1948, p. 2.

85. *NMT,* 12/11/1949, p. 2.

86. *NMT,* 8/21/1950, p. 1.

87. Janome mishin sha shi hensan iinkai, ed., *Janome mishin sōgyō 50 nen shi,* p. 392.

88. *NMT,* 8/21/1953, p. 7.

89. Nihon mishin sangyō kyōkai, *Nihon mishin sangyō shi* (Tokyo: Nihon mishin kyōkai, 1961), p. 9; Tsūshō sanyō shō, ed., *Wappu hanbai jittai chōsa*

(Tokyo: Tsūshō sanyō shō, 1962), p. 9. This survey was limited to stores willing to offer at least some goods on installment.

90. For a more detailed study of "ticket" credit, see Andrew Gordon, "From Singer to Shinpan: Consumer Credit in Modern Japan," in *The Ambivalent Consumer: Questioning Consumption in East Asia and the West,* ed. Sheldon Garon and Patricia L. Maclachlan (Ithaca, NY: Cornell University Press, 2006), pp. 137–62.

91. A useful description of the ticket business is offered by Takagi Kunio, "Chiketto hanbai no keitai," *Jurisuto,* 10/15/1967, pp. 68–69.

92. Nihon shinpan, ed., *Za bunka: Nihon shinpan no hanseiki,* p. 3; "Momeru hyakkaten no geppu uri," *Tokyo Asahi shinbun,* 11/4/1958, p. 4.

93. *Tokyo Asahi shinbun,* 10/31/1959, p. 1, and 10/24/1959, p. 4.

94. The texts of the law and Diet deliberations are available online at http://kokkai.ndl.go.jp (accessed 8/6/2010).

95. See Shinada Seihei, *Wappu hanbai no hōritsu, kaikei, zeimu* (Tokyo: Dayamondo sha, 1961), 63–218.

96. "Geppu hanbai, iijii peemento," *NMT,* 10/21/1951, p. 2.

97. Lendol Calder, *Financing the American Dream* (Princeton, NJ: Princeton University Press, 1999), pp. 166–83, 212–30; Rosa-Maria Gelpi and François Julien-Labruyère, *The History of Consumer Credit* (New York: St. Martin's Press, 2000), pp. 99–101.

98. Satō Sadakatsu, "Wappu hanbai no shidō ni tsuite," *Geppu kenkyū* 1, no. 6 (1957): p. 3; and 1, no. 8 (1957): p. 5.

99. "Kuru ka? geppu jidai," *Tokyo Asahi shinbun,* 11/21/1957, p. 5.

100. Kagawa Sanroku, "Toward Achieving Stable Economic Growth," *Geppu kenkyū* 2, no. 1 (1958): p. 2.

101. Kawauchi Mamoru, "Geppu hanbai no keizaigaku," *NMT,* 3/21/1952, p. 9.

102. "Wappu hanbai hō," *Janome shanai hō* 7, no. 37 (1962): p. 9 notes that a competitor uses this term.

103. Tokyo shōkō kaigisho, "Wappu hanbai ni kan suru jittai chōsa" (Tokyo: Tokyo shōkō kaigisho, 1957), p. 1.

104. For an example of Yajima's work, see "Wappu hanbai no keizaiteki igi," *Jurisuto,* 10/15/1957, pp. 76–78.

105. Tokyo shiyakusho, *Wappu hanbai ni kansuru chōsa* (Tokyo: Tokyo shiyakusho, 1935), p. 63.

106. Calder, *Financing the American Dream,* pp. 181–83, 217–20; on postwar United States, see also Lizabeth Cohen, *A Consumers' Republic: The Politics of Mass Consumption in Postwar America* (New York: Knopf, 2003), pp. 278–86.

107. Janome mishin sha shi hensan iinkai, ed., *Janome mishin sōgyō gojū nen shi,* pp. 246–47; *Janome shanai hō* 8, no. 51 (1963): pp. 10–11.

108. Satō Sadakatsu, "Wappu hanbai," *Geppu kenkyū* 2, no. 2 (1958): p. 4.

109. Satō Sadakatsu, "Wappu hanbai no shidō ni tsuite" *Geppu kenkyū* 1, no. 6 (1957): p. 3.

110. Satō Sadakatsu, "Wappu hanbai no shidō ni tsuite" *Geppu kenkyū* 1, no. 7 (1957): p. 2.

111. "Chiketto hanbai ni shin te: Okusan no mie wo riyō," *Tokyo Asahi shinbun,* 6/28/1954, p. 8.

112. "Kuru ka? geppu jidai," *Asahi shinbun,* 11/21/957, p. 5.

113. Satō Sadakatsu, "Wappu hanbai no shidō ni tsuite" *Geppu kenkyū* 2, no. 5 (1958): pp. 7–8; *Janome shanai hō* 7, no. 45 (1962): p. 4.

114. With the takeoff of credit cards in the 1980s and 1990s, the rhetoric of moral judgment and an image of irresponsible women shoppers have become more prominent. Young women mired in credit card debt play a prominent role, for instance, in two very popular recent mystery novels: Miyuki Miyabe, *All She Was Worth,* and Kirino Natsuo, *Out.*

115. Chalmers Johnson, *MITI and the Japanese Miracle* (Stanford, CA: Stanford University Press, 1982), p. 15.

116. Partner, *Assembled in Japan.*

117. Scott O'Bryan, *The Growth Idea: Purpose and Prosperity in Postwar Japan* (Honolulu: University of Hawaii Press, 2009).

118. Charles Maier, "The Politics of Productivity: Foundations of American International Economic Policy after World War II," in *Between Power and Plenty: Foreign Economic Policies of the Advanced Industrial States,* ed. Peter Katzenstein (Madison: University of Wisconsin Press, 1978).

119. *Saturday Evening Post,* 8/15/1959, p. 3.

7. A NATION OF DRESSMAKERS

1. Minami Hiroshi and shakai shinri kenkyūjo, eds., *Shōwa bunka, 1925–45* (Tokyo: Keisō shobō, 1987) develops this "leapfrog" interpretation in relation to daily life and dress (pp. 77, 124).

2. For this argument, see Inoue Masato, *Yōfuku to Nihonjin: Kokuminfuku to iu moodo* (Tokyo: Kōsaidō shuppan, 2001), pp. 231–33.

3. Kawasaki rōdō shi hensan iinkai, ed., *Kawasaki rōdō shi: Sengo hen* (Kawasaki: Kawasaki shi, 1987), p. 200, cites a March 1950 survey prepared by Rōdō igaku shinrigaku kenkyūjo. Similar results were reported by the Women and Minors Section of the Ministry of Labor; see Rōdō shō, fujin shōnen kyoku, *Kōjō rōdōsha seikatsu no chōsa* (Tokyo: Rōdō shō, fujin shōnen kyoku, 1952). See p. 85 for the time-use data; p. 36 for machine ownership rates.

4. For the United States, see Joan Vanek, "Keeping Busy: Time Spent in Housework, United States, 1920–1970" (PhD diss., University of Michigan, Sociology, 1973). For France, see Jean Stoetzel, "Une etude du budget-temps de la femme dans les agglomerations urbaines," *Population* 3, no. 1 (1948): pp. 52–56; and C. A. Mason, "Le budget-temps de la femmes a Londres dans les classes laborieuses," *Population* 4, no. 2 (1949), p. 372. The Mass Observation diaries were coded and analyzed by J. I. Gershuny, who kindly shared the unpublished sewing data with me. On his use of the Mass Observation data more generally, see J. I. Gershuny, "Changing Use of Time in the United Kingdom: 1937–1975, the Self-Service Era," in *Studies of Broadcasting: An International Annual of Broadcasting Science,* ed. Y. Kato and K. Goto (Tokyo: NHK, 1983), pp. 71–92.

5. Fujin shōnen kyoku, *Fujin no shokugyō ni kansuru yoron chōsa* (1954), p. 47; reprinted in *Sengo fujin rōdō seikatsu chōsa shiryōshū, Seikatsu hen,* vol. 8 (Tokyo: Kuresu shuppan, 1991), p. 240.

6. This survey was introduced in chapter 3.

7. Rōdō shō, fujin shōnen kyoku, ed., *Shufu no jiyū jikan ni kansuru ishiki chōsa* (Tokyo: Rōdō shō, fujin shōnen kyoku, 1959), pp. 2–19. Multiple answers were accepted, so the total exceeded 100 percent, and multitasking was eminently possible, especially listening to the radio in combination with any of the other choices.

8. Works on the history of the housewife's role in Western Europe and North America include Robert G. Moeller, *Protecting Motherhood: Women and the Family in the Politics of Postwar West Germany* (Berkeley and Los Angeles: University of California Press, 1993); Glenna Matthews, *"Just a Housewife": The Rise and Fall of Domesticity in America* (New York: Oxford University Press, 1987); Mary Nolan, " 'Housework Made Easy': The Taylorized Housewife in Weimar Germany's Rationalized Economy," *Feminist Studies* 16, no. 3 (1990): pp. 549–77; Karen Offen, "Body Politics: Women, Work and the Politics of Motherhood in France, 1920–1950," and Irene Stoehr, "Housework and Motherhood: Debates and Policies in the Women's Movement in Imperial Germany and the Weimar Republic," both in *Maternity and Gender Policies: Women and the Rise of the European Welfare States, 1880s–1950s,* ed. Gesela Bock and Pat Thane (London: Routledge, 1991); and Renate Bridenthal, "Professional Housewives: Stepsisters of the Women's Movement," in *When Biology Becomes Destiny: Women in Weimar and Nazi Germany,* ed. Renate Bridenthal, Anita Grossman, and Marion Kaplan (New York: Monthly Review Press, 1984).

9. Suzanne Vogel, "The Professional Housewife: The Career of Urban Middle Class Japanese Women," *Japan Interpreter* 12, no. 1 (Winter 1978): pp. 16–43.

10. For a historical definition of *sengyō,* see Kaneda Haruhiko, ed., *Gakken kokugo daijiten,* vol. 8 (Tokyo: Gakushū kenkyūsha, 1978), p. 39.

11. On the ascendance of the *sengyō shufu* as a postwar phenomenon, see Kunihiro Yōko, *Shufu to jiendaa* (Tokyo: Shōgakusha, 2001), pp. 4–6; Ochiai Emiko, *21 seiki kazoku e: Kazoku no sengo taisei no mikata, koekata* (Tokyo: Yūhikaku, 1994).

12. Vogel, "Professional Housewife," p. 17.

13. Ōya Sōichi, "Nihonjin," *Tokyo Asahi shinbun,* 1/13/1957, p. 11; cited in Nakayama Chiyo, *Nihon fujin yōsō shi* (Tokyo: Yoshikawa kōbunkan, 1987), p. 462.

14. Nakayama, *Nihon fujin yōsō shi,* pp. 455–56; Yoshimoto, "Hanabiraku yōsai gakkō," in Koizumi Kazuko, *Yōsai no jidai* (Tokyo: OM shuppan, 2004), p. 30 on U.S. goals.

15. On the New Life Movement, see Andrew Gordon, "Managing the Japanese Household: The New Life Movement in Postwar Japan," in *Gendering Modern Japanese History,* ed. Barbara Molony and Katherine Uno (Cambridge, MA: Harvard University Asia Center, 2005), pp. 423–51.

16. Inoue, *Yōfuku to Nihonjin,* pp. 236–38. Citations from the first issue of *Ifuku bunka* are on p. 237. The journal's editorial board included Narita Jun and Endō Masajirō, co-founder of the Cultural Dress Academy.

17. Yoshida Kenkichi, "*Monpe* kara A rainu made: Fukusō fūzoku no sengo 10 nen," *Bungei shunjū,* special issue, March 1955, pp. 140–42.

18. Inoue, *Yōfuku to Nihonjin,* pp. 239–40, cites Kon Wajirō, *Janpaa o kite 40 nen* (Tokyo: Bunka fukusō gakuin shuppan kyoku, 1967), p. 130.

19. "Seikatsu wa dō kaizen subeki ka?" *Yomiuri shinbun,* 8/28/1950, p. 1.

20. Nakayama, *Nihon fujin yōsō shi,* pp. 456–57, cites the editor's founding statement in *Amerikan Fuasshon,* August 1947.

21. Nakayama, *Nihon fujin yōsō shi,* pp. 458–59.

22. Yoshida, "*Monpe* kara A rainu made," p. 142.

23. Liza Crihfield Dalby, *Kimono: Fashioning Culture* (New Haven, CT: Yale University Press, 1993), p. 131.

24. "Yukata dai ninki," *Yomiuri shinbun,* 5/29/1950, evening ed., p. 2.

25. Dalby, *Kimono,* p. 131.

26. "Kane no naru ki no monogatari," *Yomiuri shinbun,* 6/7/1959, evening ed., p. 2. See also "Sakan na kimono buumu," *Yomiuri shinbun,* 12/17/1959, p. 9.

27. Dalby, *Kimono,* pp. 141–42, cites Higuchi Kiyoyuki, *Umeboshi to nihontō* (Tokyo: Non Book, 1974).

28. "Kitsuke gakkō ni natsu yasumi wa nai," *Yomiuri shinbun,* 6/30/1970, p. 18 on kimono classes; "Henshū techō," *Yomiuri shinbun,* 1/17/1963, p. 1 on frequency of wear and the "ugly *kimono* boom."

29. Nakayama, *Nihon fujin yōsō shi,* p. 456.

30. Tokyo households in 1955 numbered 1,797,000. See Nihon tōkei kyōkai, ed., *Nihon chōki tōkei sōran,* vol. 1 (Tokyo: Nihon tōkei kyōkai, 1987), p. 172. The number of dress shops in Tokyo alone at this time was comparable to the number of post offices in all of Japan. See Janet Hunter, "People and Post Offices: The Consumption of Postal Services in Japan from the Late Nineteenth Century," in *The Historical Consumer,* ed. Janet Hunter and Penelope Francks (London: Palgrave, forthcoming).

31. Nakayama, *Nihon fujin yōsō shi,* p. 461.

32. Itō Mohei, "Redi meedo taibō ron," *Fujin kōron,* May 20, 1960, p. 62.

33. Ōya Sōichi, "Nihon no kigyō: Doreme," in *Shūkan asahi* 63, no. 4 (1958): p. 36. Ōya's comparative claim is confirmed indirectly by several of the contributions in Burman, *Culture of Sewing,* all of which stress the primary role of apprenticeship, home-based instruction, or home economics classes in public schools (rather than commercial schools) as the route by which British or American women learned to sew in the twentieth century. See especially Sally I. Helventson and Margaret M. Bubolz, "Home Economics and Home Sewing in the United States, 1870–1940," in Burman, *Culture of Sewing,* pp. 304–5, 309–11.

34. Ōnuma Jun, *Bunka fukusō gakuin yonjū nen no ayumi* (Tokyo: Bunka fukusō gakuin, 1963), pp. 150–64; Nakayama, *Nihon fujin yōsō shi,* p. 462; Ōya, "Nihon no kigyō: Doreme," p. 36 on the U.S. comparison.

35. Ōya, "Nihon no kigyō: Doreme," p. 37.

36. Ibid. See also Ōya, "Nihonjin," for total of 7,000 schools with a capacity to take 500,000 students in 1957. For number of school graduates, see Sōrifu tōkei kyoku, ed., *Nihon tōkei nenkan, 1957* (Tokyo: Nihon tōkei kyōkai, 1957), pp. 436, 438–39.

37. Ōya, "Nihonjin"; Nakayama, *Nihon fujin yōsō shi,* pp. 463–64.

38. Nakayama, *Nihon fujin yōsō shi,* p. 464.

39. Yoshimoto, "Hanabiraku yōsai gakkō," p. 31.

40. Itō, "Redi meedo taibō ron," p. 62. See also Barbara Burman, "Introduc-

tion," in Burman, *Culture of Sewing,* pp. 6–8, on the importance of examining more closely the relationship between home-based and industrial production of clothing and the relatively poor reputation of ready-made wear until the 1940s in Britain and the United States.

41. *Tsurushi,* "hang," referred to clothes hanging on racks. The *–nbo* suffix was a derogatory reference to the person wearing such clothes, also used in ethnic or racial slurs.

42. Kamimura Chikako, *Nihon ni okeru senryō seisaku to josei kaihō* (Tokyo: Keisō shobō, 1992). The first director (at the behest of the American occupiers) was Yamakawa Kikue, who had been one of Japan's most important feminist thinkers and activists since the 1920s.

43. See Rōdō shō, fujin shōnen kyoku, ed., *Katei naishoku no jitsujō: Tokyo 23-ku* (Tokyo: Rōdo shō, fujin shōnen kyoku, 1955), p. 2, for the definition. Not until 1970 was homework legally regulated by the Homework Law, which defined direct selling to consumers as self-employment, or homework. The Homework Law deserves its own study. Reports on the death of a homeworker in Tokyo in 1959 due to benzol poisoning led the government to create a Homework Advisory Committee (*shingikai*), whose report led to this law.

44. Rōdō shō, fujin shōnen kyoku, ed., *Katei naishoku no jitsujō: Tokyo 23-ku,* pp. 1–3.

45. Rōdō shō, fujin shōnen kyoku, ed., *Naishoku shūgyō kihon chōsa hōkoku* (1968), pp. 5–6 for all this data.

46. Rōdō shō, rōdō kijun kyoku, *Kanai rōdō no genjō* (Tokyo: Rōdō shō, rōdō kijun kyoku, September 1976), pp. 2–5.

47. Rōdō shō, fujin shōnen kyoku, ed., *Katei naishoku no jitsujō: Tokyo 23-ku,* p. 7; *Katei naishoku no jitsujō: Osaka shi chūkan hōkoku* (1955), pp. 4–5; *Naishoku shūgyō kihon chōsa hōkoku* (1968), p. 5.

48. Rōdō shō, fujin shōnen kyoku, ed., *Katei naishoku no jitsujō: Tokyo 23-ku,* pp. 13–14; and *Katei naishoku no jitsujō: Osaka shi chūkan hōkoku,* pp. 11–13 for similar data from Osaka.

49. Rōdō shō, fujin shōnen kyoku, ed., *Naishoku shūgyō kihon chōsa hōkoku,* pp. 3, 13.

50. Rōdō shō, fujin shōnen kyoku, ed., *Katei naishoku no jitsujō: Tokyo 23-ku,* p. 3; *Katei naishoku no jitsujō: Osaka shi chūkan hōkoku,* p. 2; *Naishoku shūgyō kihon chōsa hōkoku,* pp. 8–9.

51. Rōdō shō, fujin shōnen kyoku, ed., *Katei naishoku chōsa hōkokusho: Tokyo-to bun chūkan hōkoku* (1954), p. 5; *Katei naishoku no jitsujō: Osaka shi chūkan hōkoku,* p. 3; *Naishoku shūgyō kihon chōsa hōkoku,* pp. 8–9.

52. Rōdō shō, fujin shōnen kyoku, ed., *Katei naishoku chōsa hōkokusho,* p. 5; *Katei naishoku no jitsujō: Osaka shi chūkan hōkoku* p. 3; Rōdō shō, rōdō kijun kyoku, *Kanai rōdō no genjō* (1976), p. 20; the 1976 report is not clear on whether some of the data represent mean or median totals, and might exaggerate the shift slightly, but there is no question that the relative income of the homeworkers increased compared to that of average households. The term *working household* clearly includes middle-class salarymen as well as factory workers.

53. Yoshimoto Yōko, "Onna no jiritsu o sasaeta yōsai," in Koizumi Kazuko, ed., *Yōsai no jidai: Nihon no ifuku kakumei* (Tokyo: OM shuppan, 2004), p. 71.

54. Nihon izoku kai, ed., *Ishizue: Senbotsusha izoku no taiken kiroku* (Tokyo: Nihon izoku kai jimukyoku, 1963).

55. Ibid., pp. 32–38.

56. Ibid., pp. 92–98.

57. Ibid., pp. 253–57.

58. Yoshimoto, "Onna no jiritsu o sasaeta yōsai," pp. 59–61. See also the story of N. T. from Osaka, pp. 61–62.

59. Ibid., pp. 61–62.

60. "Tatta san zen en no shihon de dekiru yōsai naishoku no tebiki," *Janome mōdo,* Spring 1962, p. 6.

61. Nakayama, *Nihon fujin yōsō shi,* p. 464.

62. Author's interview with Ohtsuki Toshiko and Takayama Kazuko, July 1, 2003.

63. Ibid.

64. "Kaimono chō," *Asahi shinbun,* 12/3/1951, evening ed., p. 3.

65. Yōsai ga dekinai de naita musume no hanashi," *Fujin kurabu,* December 1948, p. 50.

66. "Yōsaika kaisetsu ni tsuite," *Yomiuri shinbun,* 11/1/1952, p. 5.

67. Sarah A. Gordon, *"Make It Yourself": Home Sewing, Gender and Culture, 1890–1930* (New York: Columbia University Press, 2007), ch. 4 on sewing columns in American magazines. "Clotilde's Practical and Fancy Needlework" ran in the *Chicago Daily Tribune* and other papers; accessed via ProQuest Historical Newspapers.

68. "Gofujin gata wa nani o kaitai ka," *Janome shanai hō* 1, no. 2 (1956): p. 11.

69. Ibid.

70. "Janome mishin e yoserareta aijō to shinyō," *Janome shanai hō* 1, no. 2 (1956): pp. 6–7.

71. Ōya, "Nihon no kigyō: Doreme," p. 36.

72. "Hoomu yōsai jidai kitaru to iu keredo . . . ," *Josei jishin,* 3/24/1973, p. 163. On Simplicity in the United States, see Joy Spanabel Emery, "Dreams on Paper: A Story of the Commercial Pattern Industry," in Burman, *Culture of Sewing,* pp. 235–53.

73. Author's interview with Ohtsuki Toshiko and Takayama Kazuko, July 1, 2003. Emery, in "Dreams on Paper," makes no mention of any resistance to patterns among skilled home-sewers in the United States.

74. "Kon kai de heiten, *Asahi shinbun,* 4/29/1970, p. 11.

75. On the "housewife debate" (*shufu ronsō*), see Ueno Chizuko, *Shufu ronsō o yomu* (Tokyo: Keisō shobō, 1982). For a more recent capsule account, see Fujii Harue, *Sengyō shufu wa ima: Tayōka to koseika no naka de* (Tokyo: Minerubua shobō, 2002), pp. 15–22.

76. Kathleen Uno, "The Death of 'Good Wife, Wise Mother'?" in Andrew Gordon, ed., *Postwar Japan as History* (Berkeley and Los Angeles: University of California Press, 1993), pp. 294–95.

77. "Mishin wa ie de nama akubi," *Asahi shinbun,* 5/24/1969, p. 11.

78. Nihon hōsō kyōkai, ed., *Kokumin seikasu jikan chōsa* (Tokyo: Nihon hōsō kyōkai, 1970), pp. 11–12, 1172.

79. Yoshimi Shunya, *Posuto sengo shakai* (Tokyo: Iwanami shoten, 2009).

80. Mary Brinton, *Women and the Economic Miracle: Gender and Work in Postwar Japan* (Berkeley and Los Angeles: University of California Press, 1993), p. 29 for the M-curve through the 1980s and a valuable analysis of the postwar mission of the professional housewife as educator. For more recent comparative data, see the category of "labor force statistics by sex and age," online at http://stats.oecd.org/index.aspx (accessed 8/10/2009).

CONCLUSION

1. Geoffrey G. Jones and David Kiron, "Singer Sewing Machine Company: 1851–1914," Harvard Business School Case N9–804–001 (Boston: Harvard Business School Publishing, 2003), p. 1; Tim Putnam, "The Sewing Machine Comes Home," in Burman, *Culture of Sewing*, p. 269.

2. Victoria de Grazia, *Irresistible Empire: America's Advance through Twentieth-Century Europe* (Cambridge, MA: Harvard University Press, 2005).

3. Yoda Shintarō, *Gakuriteki shōryaku hō: Hanbaiin to hanbaijustu* (Tokyo: Hakubunkan, 1916), p. 5.

4. Janome mishin sha shi hensan iinkai, ed., *Janome mishin sōgyō 50 nen shi* (Tokyo: Tokiwa shoin, 1971), p. 246.

5. "Seerusuman daiichi shugi," *Janome shanai hō* 1, no. 1 (1956): p. 1.

6. Penelope Francks, "Inconspicuous Consumption: Sake, Beer, and the Birth of the Consumer in Japan," *Journal of Asian Studies* 68, no. 1 (2009): pp. 135–37, offers a useful short summary of this work, and goes on to discuss how similar developments took place in Japan.

7. Watanabe Shigeru, "Mishin kawan ka, saiken kawan ka," *Fujin kurabu* 7, no. 5 (1926): pp. 257–59.

8. Bunka fukusō gakuin, *Bunka fukusō gakuin 40 nen no ayumi* (Tokyo: Bunka fukusō gakuin, 1964), p. 4.

9. Imura Nobuko, "Kodomo no nichijō fuku o zenbu yōfuku ni," *Fujokai* 21, no. 1 (1920): 113–15.

10. Tokyo shōkō kaigisho, *Geppu hanbai seido* (Tokyo: Tokyo shōkō kaigisho, 1929), pp. 1–2.

11. *Asahi shinbun,* 1/9/1936, p. 5.

12. Judith Coffin, "Credit, Consumption, and Images of Women's Desires: Selling the Sewing Machine in Late Nineteenth-Century France," *French Historical Studies* 18 (Spring 1994): p. 783.

13. Barbara Burman, "Introduction," in Burman, *Culture of Sewing,* p. 5.

14. In general, although not in reference to this particular illustration, this is the argument of Mona Domosh, *American Commodities in an Age of Empire* (New York: Routledge, 2006), ch. 3.

15. Imura, "Kodomo no nichijō fuku o zenbu yōfuku ni," p. 113.

16. Matsunami Tadayuki, *Sugu yaki ni tatsu geppu hanbai hō* (Tokyo: Banrikaku shobō, 1930), p. 148.

17. Joan Wallach Scott, "The Mechanization of Women's Work," *Scientific American,* September 1982, p. 178.

18. Nancy Page Fernandez, "Creating Consumers: Gender, Class and the Fam-

ily Sewing Machine," in Burman, *Culture of Sewing*, p. 157. See also D. M. Douglas, "The Machine in the Parlor: A Dialectical Analysis of the Sewing Machine," *Journal of American Culture* 5, no. 1 (1982): pp. 20–29.

19. Fiona Hackney, "Making Modern Women, Stitch by Stitch: Dressmaking and Women's Magazines in Britain, 1919–39," in Burman, *Culture of Sewing*, p. 89.

20. Fernandez, "Creating Consumers," p. 166.

21. Andrew Gordon, *The Wages of Affluence* (Cambridge, MA: Harvard University Press, 1998), ch 4; William Tsutsui, *Manufacturing Ideology: Scientific Management in Twentieth-Century Japan* (Princeton, NJ: Princeton University Press, 1998), ch. 6.

22. The Modern Girl Around the World Research Group, ed., *The Modern Girl Around the World: Consumption, Modernity, and Globalization,* (Durham, NC: Duke University Press, 2008), p. 4.

23. Inoue, *Yōfuku to Nihonjin,* pp. 45–46, 48; Nakayama, *Nihon fujin yōsō shi,* p. 438.

24. This is how Nancy Page Fernandez understands the American story; see "Creating Consumers," p. 157.

25. Harry Harootunian, *Overcome by Modernity* (Princeton, NJ: Princeton University Press, 2000), explores this issue at the level of philosophy, including the philosophy of daily life as produced by Kon Wajirō (pp. 178–201). On the 1942 symposium on modernity and its raising of this point in particular, see p. 214.

26. On British and American practices, see Cheryl Buckley, "On the Margins: Theorizing the History and Significance of Making and Designing Clothes at Home"; Hackney, "Making Modern Women"; Kathryn E. Wilson, "Commodified Craft, Creative Community: Women's Vernacular Dress in Nineteenth-Century Philadelphia," all in Burman *Culture of Sewing*, pp. 55, 74, 87, 149.

APPENDIX

1. Osaka shi, shakai bu chōsa ka, ed., *Yoka seikatsu no kenkyū: Rōdō chōsa hōkoku 19* (Kyoto: Kōbundō, 1923).

2. Wendy E. Pentland, Andrew S. Harvey, M. Powell Lawton, and Mary Ann McColl, eds., *Time Use Research in the Social Sciences* (New York: Kluwer Academic/Plenum Publishers, 1999), pp. 5–8.

3. Sandor Slazai, ed., *The Use of Time: Daily Activity of Urban and Suburban Populations in Twelve Countries,* edited in collaboration with Phillip E. Converse and others (The Hague: Mouton, 1973).

4. Nihon hōsō kyōkai, ed., *Kokumin seikatsu jikan chōsa* (Tokyo: Nihon hōsō kyōkai, 1970), pp. 11–12, 1172.

5. Nihon hōsō kyōkai, *Kokumin seikatsu jikan chōsa: Hōkyū seikatsu sha, kōjō rōmu sha joshi kazoku hen* (Tokyo: Nihon hōsō kyōkai, 1943), pp. 90–93.

Select Bibliography

ARCHIVAL COLLECTIONS

Edo-Tokyo Museum
National Archives of Japan, Japan Center for Asian Historical Records, Tokyo:
    Gaimushō kiroku, "Honpō ni okeru rōdō sōgi kankei zakken: 'Shingā mishin'
    kaisha kankei."
Nature and Science Museum of the Tokyo University of Agriculture and
    Technology.
Ohara Institute for Social Research, Tokyo: Kyū-kyōchōkai shiryō
Smithsonian Institute Archives, Warshaw Collection, Washington, DC: Sewing
    Machines, box 5, folder 2
Tokyo Metropolitan Public Records Office: Gakuji kankei kenmei mokuroku
    (Tokyo fu, Tokyo shi)
United States National Archives, Washington, DC: Record Group (R/G) 331, entry
    1639, box 4840, folder 17–2, Singer Sewing Machine Company
Wisconsin State Historical Society, Madison, WI: Singer Sewing Machine Collection

JAPANESE MAGAZINES (INCLUDING YEARS CONSULTED)

Fujin kōron (1916–1970)
Fujin kurabu (1920–1950)
Fujin no tomo (1908–1940)
Fujin sekai (1906–1933)
Fujinkai (1909)
Fujo shinbun (1900–1920)
Fujokai (1910–1948)
Geppu kenkyū (1957–1966)

Mishin kōgyō (1950–1956)
Nihon mishin taimusu (*NMT*; 1946–1955)
Shakai undō tsūshin (1932–1933)
Shufu no tomo (1917–1955)

BOOKS AND ARTICLES

Abe Tsugio, ed. *Gyōkai konjaku monogatari*. Tokyo: Nihon mishin taimusu sha, 1960.

Andō Yoshinori. "Kindai Nihon ni okeru shintai no 'seijigaku' no tame ni." *Kyōiku shakaigaku kenkyū*, no. 60 (1997): 99–116.

Appadurai, Arjun. *Modernity at Large: Cultural Dimensions of Globalization*. Minneapolis: University of Minnesota Press, 1996.

Asahi shinbun, ed. *Shinbun to sensō*. Tokyo: Asahi shinbun shuppan, 2008.

Atkins, Taylor. *Blue Japan*. Durham, NC: Duke University Press, 2001.

Bayly, C. A. *The Birth of the Modern World, 1780–1914*. Oxford: Blackwell, 2004.

Benfey, Christopher. *The Great Wave: Gilded Age Misfits, Japanese Eccentrics, and the Opening of Old Japan*. New York: Random House, 2003.

Bernard, Donald R. *The Life and Times of John Manjiro*. New York: McGraw-Hill, 1992.

Breward, Christopher. "Patterns of Respectability: Publishing, Home Sewing and the Dynamics of Class and Gender, 1870–1914." In Burman, *Culture of Sewing*.

Bridenthal, Renate. "Professional Housewives: Stepsisters of the Women's Movement." In *When Biology Becomes Destiny: Women in Weimar and Nazi Germany*, ed. Renate Bridenthal, Anita Grossman, and Marion Kaplan. New York: Monthly Review Press, 1984.

Brinton, Mary. *Women and the Economic Miracle: Gender and Work in Postwar Japan*. Berkeley and Los Angeles: University of California Press, 1993.

Buckley, Cheryl. "On the Margins: Theorizing the History and Significance of Making and Designing Clothes at Home." In Burman, *Culture of Sewing*.

Burazaa kōgyō kabushiki kaisha. *Burazaa no ayumi: Sekai ni idomu*. Tokyo: Dayamondo sha, 1971.

Burman, Barbara, ed. *The Culture of Sewing: Gender, Consumption and Home Dressmaking*. Oxford: Berg, 1999.

Burman, Barbara. "Introduction." In Burman, *Culture of Sewing*.

Burman, Barbara. "Made at Home by Clever Fingers: Home Dressmaking in Edwardian England." In Burman, *Culture of Sewing*.

Calder, Lendol. *Financing the American Dream*. Princeton, NJ: Princeton University Press: 1999.

Carstensen, Fred V. *American Enterprise in Foreign Markets: Singer and International Harvester in Imperial Russia*. Chapel Hill: University of North Carolina Press, 1984.

Coffin, Judith. "Credit, Consumption, and Images of Women's Desires: Selling the Sewing Machine in Late Nineteenth-Century France." *French Historical Studies* 18 (Spring 1994): 749–83.

Cohen, Lizabeth. *A Consumers' Republic: The Politics of Mass Consumption in Postwar America*. New York: Knopf, 2003.

Cusumano, Michael A. *The Japanese Automobile Industry: Technology and Management at Nissan and Toyota*. Cambridge, MA: Harvard Council on East Asian Studies, 1985.

Dalby, Liza Crihfield. *Kimono: Fashioning Culture*. New Haven, CT: Yale University Press, 1993.

Davies, Robert Bruce. *Peacefully Working to Conquer the World: Singer Sewing Machines in Foreign Markets, 1854–1920*. New York: Arno Press, 1976.

Dayamondo sha, ed. *Mishin: Janome*. Tokyo: Dayamondo sha, 1965.

Dayamondo sha, ed. *Sekai no kigyō monogatari: Shingaa mishin*. Tokyo: Dayamondo sha, 1971.

de Grazia, Victoria. *Irresistible Empire: America's Advance through Twentieth-Century Europe*. Cambridge, MA: Harvard University Press, 2005.

Domosh, Mona. *American Commodities in an Age of Empire*. New York: Routledge, 2006.

Doreme jogakuin, ed. *Doreme saihō*. Tokyo: Doreme jogakuin, 1957.

Douglas, D. M. "The Machine in the Parlor: A Dialectical Analysis of the Sewing Machine." *Journal of American Culture* 5, no. 1 (1982): 20–29.

Dower, John W. *War without Mercy*. New York: Pantheon Books, 1986.

Embree, John. *Suyemura: A Japanese Village*. Chicago: University of Chicago Press, 1939.

Emery, Joy Spanabel. "Dreams on Paper: A Story of the Commercial Pattern Industry." In Burman, *Culture of Sewing*.

Fernandez, Nancy Page. "Creating Consumers: Gender, Class and the Family Sewing Machine." In Burman, *Culture of Sewing*.

Feuss, Harald. "Men's Place in the Women's Kingdom: New Images of Fatherhood in Taishō Japan." In *Public Spheres, Private Lives: Essays in Honor of Albert M. Craig,* ed. Gail Lee Bernstein, Andrew Gordon, and Kate Wildman Nakai. Cambridge, MA: Harvard Asia Center Monographs, 2005.

Finnane, Antonia. *Changing Clothes in China: Fashion, History, Nation*. New York: Columbia University Press, 2008.

Francks, Penelope. "Inconspicuous Consumption: Sake, Beer, and the Birth of the Consumer in Japan." *Journal of Asian Studies* 68, no. 1 (2009): 135–60.

Friedman, Walter. *Birth of A Salesman: The Transformation of Selling in America*. Cambridge, MA: Harvard University Press, 2004.

Fujii Harue. *Sengyō shufu wa ima: Tayōka to koseika no naka de*. Tokyo: Minerubua shobō, 2002.

Fujii Tadatoshi. *Kokubō fujinkai: Hinomaru to kappōgi*. Tokyo: Iwanami shoten, 1985.

Fujin shōnen kyoku. *Fujin no shokugyō ni kansuru yoron chōsa* (1954); reprinted in *Sengo fujin rōdō seikatsu chōsa shiryōshū, Seikatsu hen*, vol. 8. Tokyo: Kuresu shuppan, 1991.

Fukushima Hachirō. "Geppu, wappu, kurejitto: Sōkan 200 gō ni yosete." *Gekkan kurejitto*, no. 200 (1973): 18–24.

Gamber, Wendy. *The Female Economy: The Millinery and Dressmaking Trades, 1860–1930*. Urbana: University of Illinois Press, 1997.

Garon, Sheldon. *Molding Japanese Minds: The State in Everyday Life*. Princeton, NJ: Princeton University Press, 1997.

Gelpi, Rosa-Maria, and François Julien-Labruyère. *The History of Consumer Credit*. New York: St. Martin's Press, 2000.

Godley Andrew. "Consumer Durables and Westernization in the Middle East: The Diffusion of Singer Sewing Machines in the Ottoman Region, 1880–1930." Paper presented at the Eighth Mediterranean Social and Political Research Meeting, Florence, March 2007.

Godley, Andrew. "Homework and Sewing Machine in the British Clothing Industry, 1850–1905." In Burman, *Culture of Sewing*.

Godley, Andrew. "Selling the Sewing Machine Around the World: Singer's International Marketing Strategies, 1850–1920." *Enterprise and Society* 7, no. 2 (2006): 266–314.

Gordon, Andrew. "Business and the Corporate State: The Business Lobby and Bureaucrats on Labor, 1911–1941." In *Managing Industrial Enterprise,* ed. William Wray. Cambridge, MA: Harvard Council on East Asian Studies Monographs, 1989.

Gordon, Andrew. *The Evolution of Labor Relations in Japan: Heavy Industry, 1853–1955*. Cambridge, MA: Harvard Council on East Asian Studies Monographs, 1985.

Gordon, Andrew. "From Singer to Shinpan: Consumer Credit in Modern Japan." In *The Ambivalent Consumer: Questioning Consumption in East Asia and the West,* ed. Sheldon Garon and Patricia L. Maclachlan. Ithaca: Cornell University Press, 2006.

Gordon, Andrew. *Labor and Imperial Democracy in Japan*. Berkeley and Los Angeles: University of California Press, 1991.

Gordon, Andrew. "Managing the Japanese Household: The New Life Movement in Postwar Japan." In *Gendering Modern Japanese History,* ed. Barbara Molony and Katherine Uno. Cambridge, MA: Harvard University Asia Center, 2005.

Gordon, Andrew. *The Wages of Affluence: Labor and Management in Postwar Japan*. Cambridge, MA: Harvard University Press, 1998.

Gordon, Sarah A. *"Make It Yourself": Home Sewing, Gender and Culture, 1890–1930*. New York: Columbia University Press, 2007.

Hackney, Fiona. "Making Modern Women, Stitch by Stitch: Dressmaking and Women's Magazines in Britain, 1919–39." In Burman, *Culture of Sewing*.

Harootunian, Harry. *Overcome by Modernity*. Princeton, NJ: Princeton University Press, 2000.

Hastings, Sally A. "The Empress's New Clothes and Japanese Women, 1868–1912." *Historian* 55, no. 4 (1993): 681–82.

Hasuike Yoshiharu. "Kindai kyōikushijo ni mita jogakusei no fukusō no hensen (3)." *Kobe gakuin joshi tanki daigaku kiyō,* no. 15 (1982): 67–89.

Hata Rimuko. *Mishin saihō hitori manabi*. 3rd ed. Tokyo: Hata shoten shuppan bu, 1933.

Hata Toshiyuki. "Shingā seizō kaisha ni kan suru hōkoku" (1903). Reprinted in Matsumura Satoshi, *Kaigai jitsugyō kenshūsei hōkoku, Nōshōmushō shōkōkyoku rinji hōkoku*. Vol. 10. Tokyo: Yumani shobō, 2002.

Havens, Thomas. *Valley of Darkness: The Japanese People and World War II.* New York: W. W. Norton, 1998.

Helventson, Sally I., and Margaret M. Bubolz. "Home Economics and Home Sewing in the United States, 1870–1940." In Burman, *Culture of Sewing.*

Higuchi Tetsuko. "Waga kuni ni okeru hifuku kyōiku hatten no yōsō," *Kaseigaku zasshi* 30, no. 4 (1979): 381–86.

Hollander, Anne. *Sex and Suits: The Evolution of Modern Dress.* New York: Albert A. Knopf, 1994.

Imazu Kikumatsu. "Shingā mishin no sōgi, 1." *Rōdō kenkyū,* no. 47 (November 1951): 34–35.

Inoue Masato. *Yōfuku to Nihonjin: Kokumin fuku to iu moodo.* Tokyo: Kōsaidō shuppan, 2001.

Ishihara Osamu. *Rōdō eisei.* Tokyo: Sugiyama shoten, 1922.

Ishikawa Rokuro. *Shusse gaikōjutsu.* Tokyo: Jitsugyō no nihonsha, 1925.

Janome mishin sha shi hensan iinkai, ed. *Janome mishin sōgyō 50 nen shi.* Tokyo: Tokiwa shoin, 1971.

Johnson, Chalmers. *MITI and the Japanese Miracle.* Stanford, CA: Stanford University Press, 1982.

Jones, Geoffrey G., and David Kiron. "Singer Sewing Machine Company: 1851–1914." Harvard Business School Case N9–804–001. Boston: Harvard Business School Publishing, 2003.

Kagawa Sanroku. "Antei shita keizai kakudai no tassei." *Geppu kenkyū* 2, no. 1 (1958): 2.

Kamimura Chikako. *Nihon ni okeru senryō seisaku to josei kaihō.* Tokyo: Keisō shobō, 1992.

Kasza, Gregory J. *The State and the Mass Media in Japan, 1918–1945.* Berkeley and Los Angeles: University of California Press, 1988.

Kawasaki rōdō shi hensan iinkai, ed. *Kawasaki rōdō shi: Sengo hen.* Kawasaki: Kawasaki shi, 1987.

Kon Wajirō. *Kon Wajirō shū dai 8 kan: Fukusō shi.* Tokyo: Domesu shuppan, 1976.

Kunihiro Yōko. *Shufu to jiendaa.* Tokyo: Shōgakusha, 2001.

Kuramoto Chōji. *Atarashii gaikōjutsu.* Tokyo: Seibundō shinkō sha, 1936.

Kuwahara Tetsuya. "Shoki takokuseki kigyō no tainichi tōshi to minzoku kigyō." *Kokumin keizai zasshi* 185, no. 5 (2002): 45–64.

Kuwata Naoko. "Shimin yōsai fukyū katei ni okeru saihōka no tenkō to jirema: Narita Jun no yōsai kyōiku ron o chūshin ni." *Kyōiku gaku kenkyū* 65, no. 2 (1998): 121–30.

Kyōchōkai, ed. *Saikin no shakai undō.* Tokyo: Kyōchōkai, 1929.

Maier, Charles. "The Politics of Productivity: Foundations of American International Economic Policy after World War II." In *Between Power and Plenty: Foreign Economic Policies of the Advanced Industrial States,* ed. Peter Katzenstein. Madison: University of Wisconsin Press, 1978.

Margerum, Eileen. "The Needle as Magic Wand: Selling Sewing Lessons to American Girls After the Second World War." In Burman, *Culture of Sewing.*

Mason, Mark. *American Multinationals and Japan: The Political Economy of Japanese Capital Controls, 1899–1980.* Cambridge, MA: Harvard Council on East Asian Studies, 1992.

Matsumiya Saburo. *Sugu kiku kōkoku.* Tokyo: Mikasa shobō, 1938.

Matsunami Tadayuki. *Sugu yaku ni tatsu geppu hanbai hō.* Tokyo: Banrikaku shobō, 1930.

Matthews, Glenna. *"Just a Housewife": The Rise and Fall of Domesticity in America.* New York: Oxford University Press, 1987.

Mimura, Janice. *Planning for Empire: Reform Bureaucrats and the Japanese Wartime State, 1931–1945* (Ithaca, NY: Cornell University Press, 2011).

Minami Hiroshi, ed. *Kindai shōmin seikatsu shi.* Vol. 5. Tokyo: Sanichi shobō, 1986.

Minami Hiroshi and Shakai shinri kenkyūjo, eds. *Shōwa bunka, 1925–45.* Tokyo: Keisō shobō, 1987.

Miyashita Zenji. "Hinan ressha," In *Chichi ga kataru taiheiyō sensō: Moeru Nihon rettō,* ed. Kusuru Yoshio. Tokyo: Dōshinsha, 1969.

Modern Girl Around the World Research Group, ed. *The Modern Girl Around the World: Consumption, Modernity, and Globalization.* Durham, NC: Duke University Press, 2008.

Moeller, Robert G. *Protecting Motherhood: Women and the Family in the Politics of Postwar West Germany.* Berkeley and Los Angeles: University of California Press, 1993.

Najita, Tetsuo. *Ordinary Economies in Japan: A Historical Perspective, 1750–1950.* Berkeley and Los Angeles: University of California Press, 2009.

Nakayama Chiyo. *Nihon fujin yōsō shi.* Tokyo: Yoshikawa kōbunkan, 1987.

Nihon hōsō kyōkai. *Kokumin seikatsu jikan chōsa.* Tokyo: Nihon hōsō kyōkai, 1970.

Nihon hōsō kyōkai. *Kokumin seikatsu jikan chōsa: Hōkyū seikatsu sha, kōjō rōmu sha, joshi kazoku hen.* Tokyo: Nihon hōsō kyōkai, 1943.

Nihon hōsō kyōkai. *Kokumin seikatsu jikan chōsa: Nōgyō setai hen.* Tokyo: Nihon hōsō kyōkai, 1943.

Nihon izoku kai, ed. *Ishizue: Senbotsusha izoku no taiken kiroku.* Tokyo: Nihon hōsō kyōkai, 1963.

Nihon mishin kyōkai, ed. *Nihon mishin sangyō shi.* Tokyo: Nihon mishin kyōkai, 1961.

Nihon shinpan, ed. *Za bunka: Nihon shinpan no hanseiki.* Tokyo: Nihon shinpan keiei kikaku honbu, 1976.

Nihon tōkei kyōkai, ed. *Nihon chōki tōkei sōran.* Vol. 1. Tokyo: Nihon tōkei kyōkai, 1987.

Nishimura Ayako and Fukuda Sumiko. "Kōtō jogaku seitō no fukusō hensen ni tsuite no hito kōsatsu." *Nihon no kyōiku shigaku,* October 1989, 51–69.

Nolan, Mary. " 'Housework Made Easy': The Taylorized Housewife in Weimar Germany's Rationalized Economy." *Feminist Studies* 16, no. 3 (1990): 549–77.

Nolte, Sharon H., and Sally Ann Hastings. "The Meiji State's Policy toward Women." In *Recreating Japanese Women: 1600–1945,* ed. Gail Lee Bernstein. Berkeley and Los Angeles: University of California Press, 1991.

O'Bryan, Scott. *The Growth Idea: Purpose and Prosperity in Postwar Japan.* Honolulu: University of Hawaii Press, 2009.

Ochiai Emiko. *21 seiki kazoku e: Kazoku no sengo taisei no mikata, koekata.* Tokyo: Yūhikaku, 1994.

Oddy, Nicholas. "Beautiful Ornament in the Parlour or Boudoir: The Domestication of the Sewing Machine." In Burman, *Culture of Sewing.*

Offen, Karen. "Body Politics: Women, Work and the Politics of Motherhood in France, 1920–1950." In *Maternity and Gender Policies: Women and the Rise of the European Welfare States, 1880s-1950s,* ed. Gesela Bock and Pat Thane. London: Routledge, 1991.

Ōnuma Jun. *Bunka fukusō gakuin yonjū nen no ayumi.* Tokyo: Bunka fukusō gakuin, 1963.

Osaka shi, shakai bu chōsa ka, ed. *Yoka seikatsu no kenkyū: Rōdō chōsa hōkoku 19.* Kyoto: Kōbundō, 1923.

Partner, Simon. *Assembled in Japan: Electrical Goods and the Making of the Japanese Consumer.* Berkeley and Los Angeles: University of California Press, 1999.

Putnam, Tim. "The Sewing Machine Comes Home." In Burman, *Culture of Sewing.*

Reynolds, Helen. " 'Your Clothes Are Materials of War': The British Government Promotion of Home Sewing During the Second World War." In Burman, *Culture of Sewing.*

Rōdō shō. *Kanai rōdō ni kan suru chōsa.* Tokyo: Rōdō shō, 1975.

Rōdō shō, fujin shōnen kyoku, ed. *Katei naishoku chōsa hōkokusho: Tokyo-to bun chūkan hōkoku.* Tokyo: Rōdō shō, fujin shōnen kyoku, 1954.

Rōdō shō, fujin shōnen kyoku, ed. *Katei naishoku no jitsujō: Osaka shi chūkan hōkoku.* Tokyo: Rōdō shō, fujin shōnen kyoku, 1955.

Rōdō shō, fujin shōnen kyoku, ed. *Katei naishoku no jitsujō: Tokyo 23-ku.* Tokyo: Rōdō shō, fujin shōnen kyoku, 1955.

Rōdō shō, fujin shōnen kyoku, ed. *Kōjō rōdōsha seikatsu no chōsa.* Tokyo: Rōdō shō, fujin shōnen kyoku, 1952.

Rōdō shō, fujin shōnen kyoku, ed. *Naishoku shūgyō kihon chōsa hōkoku.* Tokyo: Rōdō shō, fujin shōnen kyoku, 1968.

Rōdō shō, fujin shōnen kyoku, ed. *Shufu no jiyū jikan ni kansuru ishiki chōsa.* Tokyo: Rōdō shō, fujin shōnen kyoku, 1959.

Rōdō shō, rōdō kijun kyoku. *Kanai rōdō no genjō.* Tokyo: Rōdō shō, rōdō kijun kyoku, 1976.

Sakai Nobuko. "Kinoshita Takeji to saihō gakushū ron." *Katei kagaku,* no. 76 (1978): 44–54.

Sand, Jordan. *House and Home in Modern Japan: Architecture, Domestic Space, and Bourgeois Culture, 1880–1930.* Cambridge, MA: Harvard Asia Center Monographs, 2003.

Sato, Barbara. *The New Japanese Woman.* Durham, NC: Duke University Press, 2003.

Scott, Joan Wallach. "The Mechanization of Women's Work." *Scientific American,* September 1982, 166–87.

Seligman, E. R. A. *The Economics of Installment Selling.* Vol. 1. New York: Harper and Brothers, 1927.

Shimazono Satoshi. *Dotō o koete: Yamamoto Tōsaku denki.* Tokyo: privately published, 1960.

Shimizu Masami. *Hōmon hanbai chūmon o toru hiketsu.* Tokyo: Clark sōsho kankōkai, 1924.

Shimizu Masami. *Shin gaikō hanbai jutsu.* Tokyo: Seibundō shinkō sha, 1937.

Shimotani Masahiro. "The Formation of Distribution *Keiretsu:* The Case of Matsushita Electric." In *The Origins of Japanese Industrial Power,* ed. Etsuo Abe and Robert Fitzgerald. London: Frank Cass, 1995.

Shinada Seihei. *Wappu hanbai no hōritsu, kaikei, zeimu.* Tokyo: Dayamondo sha, 1961.

Silverberg, Miriam. *Erotic Grotesque Nonsense: The Mass Culture of Japanese Modern Times.* Berkeley and Los Angeles: University of California Press, 2006.

Souda Satoru, *Saihō mishin shiyō hō zensho.* Tokyo: Souda shuppan bu, 1930.

Stoehr, Irene. "Housework and Motherhood: Debates and Policies in the Women's Movement in Imperial Germany and the Weimar Republic." In *Maternity and Gender Policies: Women and the Rise of the European Welfare States, 1880s–1950s,* ed. Gesela Bock and Pat Thane. London: Routledge, 1991.

Suzuki Jun. *Shin gijutsu no shakai shi.* Tokyo: Chūō kōron sha, 1999.

Takagi Kunio. "Chiketto hanbai no keitai," *Jurisuto,* 10/15/1967, 68–69.

Takahashi Keiji. *Fujin katei naishoku.* Tokyo: Seikadō, 1919.

Tōbu teishin kyoku. *Naishoku shōrei ni kan suru jikō.* Tokyo: Teishin kyoku, 1915.

Tokyo shi, shakai kyoku. *Naishoku ni kansuru chōsa.* Tokyo: Tokyo shi, shakai kyoku, 1926.

Tokyo shiyakusho, ed. *Wappu hanbai ni kansuru chōsa.* Tokyo: Tokyo shiyakusho, 1935.

Tokyo shōkō kaigisho, ed. *Geppu hanbai seido.* Tokyo: Tokyo shōkō kaigisho, 1929.

Tsūshō sangyō shō, ed. *Wappu hanbai jittai chōsa.* Tokyo: Tsūshō sangyō shō, 1962.

Tsutsui, William. *Manufacturing Ideology: Scientific Management in Twentieth-Century Japan.* Princeton, NJ: Princeton University Press, 1998.

Ueno Chizuko. *Shufu ronsō o yomu.* Tokyo: Keisō shobō, 1982.

Uno, Kathleen. "The Death of 'Good Wife, Wise Mother'?" In *Postwar Japan as History,* ed. Andrew Gordon. Berkeley and Los Angeles: University of California Press, 1993.

Ushigome Chie. "Saihō kyōiku kaisō 50 nen." *Gakuen* (Shōwa joshi daigaku), no. 245 (July 1960): 22–38.

Vogel, Suzanne. "The Professional Housewife: The Career of Urban Middle-Class Japanese Women." *Japan Interpreter* 12, no. 1 (1978): 16–43.

Wickramasinghe, Nira. "The Reception of the Singer Sewing Machine in Colonial Ceylon/Sri Lanka." Unpublished paper presented at Princeton University, Davis Center, March 27, 2009.

Wilson, Kathryn E. "Commodified Craft, Creative Community: Women's Vernacular Dress in Nineteenth-Century Philadelphia." In Burman, *Culture of Sewing.*

Wilson, Verity. "Dressing for Leadership in China: Wives and Husbands in an Age of Revolutions." In *Material Strategies: Dress and Gender in Historical Perspective,* ed. Barbara Burman and Carol Turbin. Malden, MA: Blackwell, 2003.

Yoda Shintarō. *Gakuriteki shōryaku hō: Hanbaiin to hanbaijustu.* Tokyo: Hakubunkan, 1916.

Yōfuku kisha kurabu, ed. *Nihon no yōfuku shi.* Tokyo: Yōfuku kisha kurabu, 1976.

Yoshida Gen. "Nihon saihō mishin shi zakkō." *Mishin sangyō,* no. 100 (1967): 1–10.

Yoshimoto Yōko. "Hanabiraku yōsai gakkō." In *Yōsai no jidai,* ed. Koizumi Kazuko. Tokyo: OM shuppan, 2004.

Yoshimoto Yōko. "Onna no jiritsu o sasaeta yōsai." In *Yōsai no jidai: Nihon no ifuku kakumei,* ed. Koizumi Kazuko. Tokyo: OM shuppan, 2004.

Young, Louise. *Japan's Total Empire: Manchuria and the Culture of Wartime Imperialism.* Berkeley and Los Angeles: University of California Press, 1998.

Index

TEXT
10/13 Sabon

DISPLAY
Sabon

COMPOSITOR
Integrated Composition Systems